# Palgrave Studies in Economic History

**Series Editor**
Kent Deng
London School of Economics
London, UK

Palgrave Studies in Economic History is designed to illuminate and enrich our understanding of economies and economic phenomena of the past. The series covers a vast range of topics including financial history, labour history, development economics, commercialisation, urbanisation, industrialisation, modernisation, globalisation, and changes in world economic orders.

More information about this series at
http://www.palgrave.com/gp/series/14632

David Pretel

# Institutionalising Patents in Nineteenth-Century Spain

palgrave
macmillan

David Pretel
El Colegio de Mexico (COLMEX)
Mexico City, Mexico

Palgrave Studies in Economic History
ISBN 978-3-319-96297-9          ISBN 978-3-319-96298-6    (eBook)
https://doi.org/10.1007/978-3-319-96298-6

Library of Congress Control Number: 2018951444

This Palgrave Pivot imprint is published by the registered company Springer Nature Switzerland AG
The registered company address is: Gewerbestrasse 11, 6330 Cham, Switzerland

# Praise for *Institutionalising Patents in Nineteenth-Century Spain*

"Pretel's brilliant and concise book explores the Spanish patent system during the second half of the 19th century. He shows how this system was in the hands of intermediaries, lawyers and consulting engineers, who provided a complete response to the necessities of companies in Spain and abroad, particularly large foreign industrial firms. The Spanish patent system was planned and organised to stimulate technology transfer and the dissemination of technologies from France, the UK and the USA as a response to the perceived industrial underdevelopment of the Spanish economy."

—Santiago M. López, *University of Salamanca, Spain*

"David Pretel's book offers an original and stimulating analysis of the ambiguous role of patents in Spanish economic history. It subtly highlights the contradictions of the system by looking at its practical implementation and at players as essential as patent agents. His approach at different scales makes his work a significant contribution to the history of globalisation."

—Gabriel Galvez-Behar, *University of Lille, France*

"Pretel provides in his book a critical analysis of the Spanish patent system in the nineteenth century. It offers a nuanced and intelligent study of the legal and administrative institutions linked to technology in metropolitan Spain and its colonies. The book contains a coherent narrative that provides new outlooks on the interplay among institutions, cultural and political change, technological development, and economic growth in peripheral countries of the Atlantic space before the Great War."

—Juan Pan-Montojo, *Autonomous University of Madrid, Spain*

# Acknowledgements

I am deeply grateful to those who made the writing of this book possible. Patricio Sáiz gave generous guidance and astute comments back in the days when I was developing work on this topic as part of my doctoral dissertation at Universidad Autónoma de Madrid. During my years at the University of Cambridge, the advice of Martin Daunton was crucial to the completion of the second part of my doctoral thesis. This laid the foundation for the research on which this book is chiefly based. I am particularly grateful to Ian Inkster for guiding my thinking and encouraging my scholarship throughout the years. His wisdom and mentoring have been invaluable.

This book has also benefited enormously, in one way or another, from the comments and suggestions of Elizabeth Álvarez, Lino Camprubí, Francisco Cayón, Kent Deng, Nadia Fernández de Pinedo, Reinaldo Funes, Anna Guagnini, Adrian Leonard, Santiago López, Christine MacLeod, Alessandro Nuvolari, Naomi Lamoreaux, Oscar Pretel, Martin Rodrigo, Nivardo Silva, Suzanne Smith and María Cecilia Zuleta. I would also like to thank Palgrave's editorial team, especially Clara Heathcock and Laura Pacey, for their support and patience.

The writing of this book was supported by funding from the Spanish Ministerio de Economia y Competitividad, Project HAR2016-75010-R (La Desigualdad Económica en la España Contemporánea).

# Contents

# Abbreviations

| | |
|---|---|
| AHN | Archivo Histórico Nacional, Madrid |
| AHOEPM | Archivo Histórico de la Oficina de Patentes y Marcas, Madrid |
| ANC | Archivo Nacional de Cuba, La Habana |
| BN | Biblioteca Nacional, Madrid |
| BOPI | Boletín Oficial de la Propiedad Industrial |
| CLE | Colección Legislativa de España |
| ELZ | Elzaburu Agency Private Records, Madrid |
| TAEP | The Thomas Alva Edison Papers, Rutgers |
| TCIPA | Transactions of the Chartered Institute of Patent Agents |
| WIPO | World Intellectual Property Organisation |

# List of Figures

# List of Tables

# 1

# Institutionalising Backwardness

**Abstract** The relative industrial backwardness of nineteenth-century Spain set the stage for the institutionalisation of its patent system between 1826 and 1902. International patterns of technological development, industrialisation and trade largely explain the distinctive peripheral character of Spain's patent system and its institutional organisation. The institutionalisation of patent rights in this country went hand in hand with the expansion of a new industrial culture, socio-technical infrastructure and a rhetoric in support of intellectual property rights. This chapter provides the historical context and analytical framework for an understanding of the evolution of the Spanish patent institution and culture during the nineteenth century.

**Keywords** Backwardness • Technology • Culture • Institutions • Agency • Patents

On 1 May 1851, Queen Victoria inaugurated The Great Exhibition of the Works of Industry of all Nations in London, hailed as an unprecedented global event. The Exhibition took place in Hyde Park in a single sizeable

© The Author(s) 2018                                                               1
D. Pretel, *Institutionalising Patents in Nineteenth-Century Spain*, Palgrave Studies in Economic History, https://doi.org/10.1007/978-3-319-96298-6_1

landmark building that would become an icon of modernity, the Crystal Palace designed by Joseph Paxton. In record time, only 11 months, this massive prefabricated building would be constructed with vast quantities of steel, timber and glass supplied by manufacturers from several parts of the country. The pioneering introduction of the railway throughout Britain provided the infrastructure to move the various components to London. The large exhibition hall that housed this great exaltation and celebration of modern industry was in itself the foremost example of British economic hegemony during the Victorian era. Great Britain was by then the first industrial country, the factory of the world. Inside the Crystal Palace, dozens of countries, including Spain, exhibited their most representative products, from steam engines to samples of wool or charcoal to extravagant art samples such as the Medieval Court of the architect A. W. N. Pugin.

According to the chronicles of the time, the massive scale of the steam engines exhibited in the Machinery Hall sparked great admiration among the over six million visitors who attended London's Exhibition. The celebration of inventions and inventors was an essential part of the event. Technology was seen as the source of British economic superiority, and the Great Exhibition was a cathedral of industrial progress. Concerns about the social consequences of mechanisation and the relative decline of British industrialisation were, however, not uncommon during the Great Exhibition as evidenced by the opinions of, among others, Charles Babbage and Lyon Playfair.[1]

In addition to highlighting British industrial supremacy and economic power, the London Exhibition served as a space to unleash international comparisons. The public presentation of the latest mechanical inventions before the delegates of various countries itself had an exacerbating effect on nationalism. Periodicals and popular writings on the Exhibition made stereotyped judgements between the civilised West and the barbarian East, between prosperous Protestantism and impoverished Catholicism. While the internationalist ideal was one of the pillars of the event, the apparent material disparities that emerged during the Exhibition led local commentators to emphasise British exceptionalism.[2]

At the Crystal Palace, each country was assigned a space to exhibit its cutting-edge products, techniques and designs. Nearly half of the approximately 14,000 total exhibitors were from Great Britain. Spain had 289 exhibitors (see Fig. 1.1 for an illustration of the kinds of products

**Fig. 1.1** Spain's exhibit at London's Crystal Palace Exhibition in 1851. (Source: Dickinson's Comprehensive pictures of the Great Exhibition of 1851: from the originals painted for H.R.H. Prince Albert (London: Dickinson Brothers, 1854))

showcased by Spanish manufacturers and artisans). The chronicles of the Exhibition evidenced the limited relevance and quality of the goods and inventions presented by Spain, while the official catalogue of the Exhibition directly deplored Spain's dismal participation.[3] Catalonia, despite being that country's most industrially advanced region, only contributed 24 exhibitors to the London World's fair. One of the Spanish representatives at the Exhibition, the Catalan economist and politician Joan Yllas Vidal, admitted that he blushed with disappointment when he visited the Spanish section.[4] Yllas wrote extensively about this world's fair, including his *A Look at the London Exhibition* published in Barcelona in 1852.[5] The Spanish magazine *La Ilustración* went a step further and correlated British mechanical superiority with the reliability of its patent system, which encouraged invention and industrial progress.[6] In a similar vein, the weekly *The Illustrated London News* also drew attention to the poor Spanish participation, despite Spain's superior natural resources, and blamed Spain's lack of industrial wealth to its pervasive traditions and prejudices, which the magazine claimed were more influential than its laws.[7]

Despite the liberal reforms introduced in Spain during the two decades prior to the 1851 world's fair, the country remained technologically backward relative to industrial leaders, particularly Britain. A modern patent law had been introduced, engineering schools had been established and tariffs on machinery imports had been reduced—all with the goal of promoting technological progress, yet without much success. Although some new industrial sites, expert professionals and mechanical journals had emerged, Spain primarily remained an impoverished, agrarian and illiterate country, with very limited technological stock and industrial capabilities. This relative industrial failure was explained by many contemporary commentators as the consequence of Spanish culture and institutions.[8] According to the stereotype, the essence of Spanish culture was a proclivity for leisure, a lack of entrepreneurial initiative and a disregard for the law. This characterisation contrasted sharply with the popular image of mid-nineteenth-century British culture, which was identified with creativity, rationality, innovation, moderation and openness to change.[9]

Yet while Spain was far from enjoying a widespread use of industrial technologies, the machine gradually became the dominant symbol of economic progress in this country during the mid-nineteenth century. Foreign technologies became signifiers of modernity while national artisanal products were increasingly considered archaic. Notwithstanding Spain's apparent industrial backwardness, domestic political and economic elites were attracted—with important exceptions such as some agrarian classes—by the virtues of mechanisation and its promise of a better material future.[10] In his 1865 article 'Should Spain Be an Industrial Country?' Ramón de Morenés, deputy director of the Telegraph Corps, who would win a bronze medal at the Universal Exhibition of Paris in 1867, put it eloquently: 'When industry develops, the products of the earth will be multiplied a hundredfold and the most painful tasks will be done efficiently and cheaply by powerful iron machines'.[11] Along similar lines, but some years later, the politician and jurist Manuel Dánvila, instrumental in regulating intellectual property rights in Spain during the 1870s, declared in 1882 that the 'civilising action of steam and electricity' demanded the universal protection of human inventiveness and industrial activity.[12]

## Spain's Technological Backwardness

In the European context, Spain was a paradigmatic example of relatively low levels of industrial innovation during the nineteenth century.[13] This may explain why Spain, despite periods of considerable growth, failed to achieve any real convergence, in macroeconomic terms, with the most advanced European economies during the second half of the nineteenth century.[14] The history of nineteenth-century Spanish industrialisation is one of assimilation of already available foreign technologies. It is not one of invention but of imitation and adaptation to local contexts. In other words, it is the history of a delayed mechanisation and industrialisation of a technologically backward country in the European periphery. Over the course of the nineteenth century, technical capabilities and industrial productivity remained lower in Spain than on the other side of the Pyrenees.

Although it had failed in its attempts at industrialisation, nineteenth-century Spain was not without industry and machines.[15] The introduction of industrial technology followed well-defined regional patterns and was concentrated in a range of sectors, including metallurgy, textiles, railways, weapons, chemicals and commercial services. Most Spanish regions failed to industrialise; others, such as Catalonia, were early-industrialisers (thanks notably to Barcelona's cotton textile industry).[16] Likewise, some Spanish regions of the maritime periphery and the capital city of Madrid would follow a pattern of late industrialisation during the final decades of the nineteenth century. Several industrial districts in smaller towns in Catalonia, Valencia and the Basque Country would also concentrate sector-specific technological microcultures with higher innovation rates during the late nineteenth and early twentieth centuries.[17]

Even when some Spanish regions did industrialise, relatively speaking, they remained well below the production levels of the most advanced industrial regions of Western Europe and the United States. In general, the last two decades of the nineteenth century in Spain were a period of economic depression and failed economic convergence with Western Europe and the United States. By the turn of the century, Spain was still not an industrial country and had lost its colonial possessions in the Caribbean and the Pacific. The scope of Spain's lagging development, industrial failures and political disasters during the Restoration regime of the late nineteenth century has remained a controversial question in the historiography for decades, despite abundant empirical data.[18]

In an increasing international context of economic globalisation and integration of the world markets, nineteenth-century Spain remained dependent on European technologies in developing its national industries. This happened across all sectors. Spain's timid industrialisation was built on the importation of technology in an effort to benefit from technological transfer without incurring the costs of promoting original invention. Spain and its extra-European territories (Cuba, Puerto Rico and the Philippines) were, in effect, colonised by foreign technologies and failed to develop a machine-making manufacturing base of their own. It was not until the last two decades of the nineteenth century that Spain began to introduce a significant degree of modern industrial technologies—though without reaching the levels of the leading countries. But, once again, the steady Spanish technological advancements during the later part of the century were largely built on foreign technologies and expertise.

During the second half of the nineteenth century, the diffusion of manufacturing technologies and processes appears to have been highly concentrated in the urban areas of Bilbao, Barcelona, Madrid and Málaga, where the industrial working and middle classes developed.[19] These cities saw the emergence of industrial sites that constituted dynamic and advanced clusters where engineers and businesspeople were aware of the latest innovations and participated in international networks of technological knowledge exchange. For instance, in 1861 almost half of the steam engines installed in Spain for industrial purposes went to the

Catalan cotton industry, mainly in Barcelona. That year, virtually all Catalan spinning and 45% of Catalan weaving were mechanised.[20]

Not only did Spain experience a pattern of technological change different from that of early industrialised countries, such as Britain, but it also represented a historical case that differed significantly from latecomers such as Germany, Italy or Japan. Although some of these latter nations were also highly dependent on existing foreign technology, they managed to develop domestic innovative activity in several economic sectors by the late nineteenth century. The increasing Spanish dependence on foreign products and technologies from the 1860s may explain the decision of the government to raise tariffs in 1891. Some Spanish industrial producers, such as the machine-builder La Maquinista Terrestre, had lobbied during the 1880s to obtain greater commercial protection for Spanish mechanical production.[21]

Even when the general pattern was one of relatively slow technological change, the historiography on nineteenth-century Spain has identified several examples of an early introduction of foreign advanced technology to the country, chiefly the textile manufacturing industry in Catalonia and the Basque iron and heavy industries. Several electrical, agricultural, railway and shipping technologies were also introduced relatively early to certain Spanish regions.[22] Yet, despite some successful attempts, many of the episodes of technology transfer to Spain during the mid-nineteenth century failed to be adopted or widely diffused. The late nineteenth century witnessed a further divergence between Spain's technological capacities and that of successful latecomers such as Germany and the United States. The envisioned transition from foreign to domestic technological capabilities, in short, had failed.

# Explaining Backwardness: Culture and Institutions

Historians and economists have long discussed several possible explanations for Spanish technological backwardness during the nineteenth century. A popular explanation invokes an analysis of its broad national culture and related cultural variables. Possible cultural inhibitors range from the behaviours of the country's politicians to the attitudes of its entrepreneurs. For instance, historians have pointed to a proclivity among

the Spanish liberal bourgeoisie and business classes for technological conservatism, an aversion to risk, a resistance (or at least an indifference) to
machinery investment and a reluctance to compete in international markets.[23] While cultural and entrepreneurial explanations for Spanish industrial backwardness contain an element of truth, they may have been
exaggerated and have been insufficient to explain Spain's fate in the
Second Industrial Revolution. Some Spanish businessmen exhibited a
more positive, receptive and welcoming attitude to technology transfer
than one might expect. Nineteenth-century Spain was not a homoge-
neous culture, nor did it have a simple cultural trajectory; some environments were far more innovative than others, and some businessmen were
more welcoming to mechanical investments than others. Moreover, businessmen in all societies, including the most industrially advanced, tended
towards conservative behaviours and had an aversion to risk; highly innovative and productive entrepreneurs were a minority.[24] The idea that
innovation stemmed primarily from entrepreneurs' attitudes is vague and
fraught with problems. The national cultural traits affecting innovation
are difficult to isolate and their direct influence difficult to demonstrate.

Not only were entrepreneurs a necessary element in Spanish industrialisation, but they were also a result of the industrialisation process itself.
Where businessmen chose to allocate their resources depended to a large
extent on the economic incentives to innovation. The relationship between
attitudes towards technology and the degree of innovation is not direct; it
is heavily influenced by economic imperatives. Technological change and
innovative entrepreneurship depended on the structure of rewards and the
opportunities presented by the Spanish economy during the nineteenth
century. It seems that the crucial factor here was not only the nature of
Spanish culture but rather the relatively small dimension of the country's
urban and industrial sites as well as the limited extension of its middle and
business classes. For example, the low degree of integration of the Spanish
market and the limited demand from downstream industries (such as shipbuilding, metal transformation and railways) did not incentivise technical
innovation in the country's metallurgical industries. Technologies require
an initial investment and a capital commitment, both of which constrain
the adoption of these expensive and complex technologies when the level of
demand and the scale of production have not reached a certain minimum.

Technological change cannot be reduced to economic determinants and cultural constraints. Productive as they may be, a narrow economic explanation or a broad cultural approach does not seem to provide sufficient explanations for Spain's technological backwardness. The answer must be sought in the synergy of a number of non-mutually exclusive explanations rather than a single one. Economic, demographic and geographic explanations should be conjoined with institutional explanations that take into account the changing nature of technological cultures in particular innovative sites. While culture broadly speaking is essential in understanding nineteenth-century technological change, the culture of innovation is more identifiable when delimited and located in precise systems and institutions. A single industrial site might express multiple cultural identities, encompassing both a willingness to introduce complex foreign industrial technologies and to engage in rent-seeking behaviours. Time is also important. Technological cultures transformed and adapted to the global context during the nineteenth century, as they faced ever-increasing pressures for homogenisation.

The historical study of institutions can provide an additional, more nuanced and less deterministic explanation for Spain's nineteenth-century technological history. Culture and institutions are two sides of the same coin. Institutions are the receptacles of collective beliefs and values, but also of material life and the past. The institutional approach, however, is also problematic. Assuming that institutions play a central role in economic development, several economic historians and economists have related failed industrialisation to inadequate economic institutions, assuming these to be a primary manifestation of national culture.[25] From this liberal-conventional perspective, inefficient institutions are a barrier to technological progress. The problem with this approach is that while it rightly recognises the critical role of institutions, it often does not pay enough attention to the actual historical making of, and agents involved in, such institutions. Culture is embodied in institutions, but institutional change is driven by politics, social agency and decisions rather than by broad cultural traits understood in their Weberian longue durée sense.

Institutions should be historicised rather than taken as exogenous variables. From an institutionalist perspective, institutions—whether formal or informal—are not immutable or simply natural. Ultimately, institutions—such as property rights—have been redefined in diverse ways in varying temporal and social contexts.[26] Institutions are much a result of economic development as a cause.[27] Therefore, the analysis of institutions should avoid broad economic and cultural determinism wherein politics and social agency are exogenous variables and should instead unpack institutional organisation in precise historical conjunctures.

The historical study of institutional imperatives for late industrial development advanced by Alexander Gerschenkron in 1962—including the analysis of the national institutions that facilitated technological change and transfer—should provide a sufficient explanation for Spain's relative technological backwardness during the second half of the nineteenth century.[28] In latecomers, state-driven institutional innovation could facilitate rapid catching-up and even leadership in new advanced industries. However, the study of national imperatives should be complemented by a consideration of international dynamics and non-national actors. Complex institutions of the state, such as patent systems, nurtured a delimited socio-technical microculture that cannot be explained merely by the broader national culture but also arose from inner institutional innovations and foreign influences.[29]

At the same time, the character of new technologies was transforming the imperatives of late industrialisation. Although they are two interconnected processes, there was a discontinuity between the dynamics of technological change of the first industrialisation and those of the Second Industrial Revolution. From the 1870s, as the economic historian Joel Mokyr puts it, 'the history of technology becomes a different tale'.[30] Around this decade there was a gradual transition to a new technological paradigm. The late nineteenth century was the age of the rise of electrical and chemical technologies, automatic mechanical equipment and tools, special purpose technologies for mass production, manufacturing machinery with interchangeable parts, the internal combustion engine, large-scale communication systems and massive transport infrastructures. These technological innovations were institutionally embedded in particular industrial sites and usually subject to intellectual property rights, professional research and standardised regimes. The classic example of this transition is the development of the German chemical industry in the 1860s. Other imperatives associated with new electromechanical and chemical technologies were large capital investments, professional expertise and the supply of raw materials from far-flung commodity frontiers. These new requirements were concomitant with the gradual rise of large science-based industries and multinational companies in a number of mass-production sectors.

# Technology, Patents and Agency

A critical history of technology in nineteenth-century Spain requires attention not only to the conditions of backwardness but also to knowledge structures, such as institutional systems. The nineteenth-century Spanish liberal revolution and modernisation project—although not wholly satisfactory—advanced significant institutional changes in an effort to foster technological progress.[31] One of the reforms that Spanish economic and political elites regarded as an institutional imperative for industrialisation was the regulation of intellectual property rights, particularly patents. The study of the formal institution conferring patent rights in Spain can provide an empirical case study of a central knowledge system during the nineteenth century.

Patent systems are institutional innovations that proliferated in the Atlantic nations throughout the nineteenth century. An important attribute of technology is its ownership, as it affects its production, distribution and use. Patent institutions are not only systems granting monopolies but—as standardised repositories that recorded appropriated technological knowledge—information systems. This latter attribute makes patents a formal mechanism for the international circulation of technology. Especially from the late nineteenth century, the growing interrelation among national patent systems went hand in hand with the globalisation of production and trade. International patenting frequently involved not just the international transfer of private rights, and eventually technologies, but also the transmission of ideas, perceptions and information.[32]

Defining technology has long been a central epistemological problem. Technology is a form of knowledge and as such its study should take into consideration not only its embodiment as a machine or organisational process but also the practices, actors, spaces and institutions of knowledge production, circulation and use.[33] In particular, the study of patent institutions poses interesting questions about institutional agency. If patent institutions are knowledge structures with a socio-cultural history, then it follows that the understanding of patent institutions requires attention to the 'players', that is, those using and shaping the system, such as the patentees and regulators.[34] However, any patent system is more than just inventors and patent offices. When studying patent cultures, other actors should be examined as well. Especially relevant are those social actors who occupied some intermediate position between patentees, manufacturers, investors and the patent office. Also, core is those actors, from industrialists to engineers, acting as partners of patentees.

Institutional diversity is another central issue when studying patent systems.[35] The essential insight here is that although the various nineteenth-century national patent institutions were allegedly analogous and built following similar principles, the existing patent institutions had diverse purposes and were institutionalised in distinct ways. As the economist Paul A. David claimed, and several historians have demonstrated more recently, intellectual property rights have throughout history been redefined in practical ways to accommodate national interests and changing views.[36] Access to information, the cost of patenting, bureaucracy, patentability, technical examination, legal enforcement and market structure were just some of the factors that varied from one national context to another, specifically from the core industrial nations to the peripheral latecomers. The heterogeneous institutional evolution of patent systems responded to both the socio-economic context of each society and to the determinants of the international economy. In this sense, patent systems were not just an institution for the protection of the rights of inventors, but an instrument of government intervention for the promotion of national or local economies and interests.[37]

Patent rights were and remain a central subject of public policy debate. Indeed, the role of intellectual property rights in facilitating economic growth is as much a subject of criticisms nowadays as it was during the European patent controversy of the mid-nineteenth century. Intellectual property rights lay at the heart of the history of industrialisation, yet they introduce a major contradiction into capitalism as they artificially create knowledge scarcity and reduce competition through market intervention.[38] In a strict sense, technology is a public good, and hence any monopoly on the utilisation of technology has some social externalities. The temporary declaration of private ownership on inventions has been seen by many economists and governments as the best policy (even an institutional prerequisite) by which to stimulate innovation. Economists such as Kenneth Arrow and Douglass North, among others, have pointed out the importance of secure patent rights to economic development.[39] From this perspective, clearly defined intellectual property rights provide superior and efficient incentives to invention and are a key instrument with which to induce innovation. The rationale for patent rights is that institutions under the authority of states ultimately stand in for market forces. Through temporary monopolies, inventors can protect and profit from their technological inventions, while society benefits from the disclosure of technological knowledge and the innovation that ensues.

Doubts have, however, been expressed about the historical role of patents in fostering technological progress.[40] Empirical evidence has shown that patent rights were not always an effective instrument with which to stimulate innovation nor the prime channel for the dispersal of knowledge during the nineteenth century.[41] Take, for instance, the knowledge sharing through collective invention arrangements, prizes for technical improvements and mechanical exhibitions at world fairs.[42] During the twentieth and twenty-first centuries, patents have created further dissatisfaction among developing economies, as granting exclusive rights to inventors creates legal restrictions on the international diffusion of potentially valuable technological knowledge.[43] The suitability of patent institutions for developing nations is indeed a controversial matter, as is any monopolistic policy that restricts or slows down the spread of, or access to, technologies. In the words of David Harvey, 'technological systems have been used not only to protect monopoly powers but also to secure accumulation by dispossession'.[44] This criticism has increased in recent years as the domain of patent protection has been expanded to include, among other things, indigenous peoples' knowledge, genetic materials and biological products.[45] Similarly, a critique of patent rights immediately raises questions about the necessary balance between the interests of patent holders and the society at large, that is, between the private and the public. Predatory strategies by lawyers and multinationals and costly judiciary systems—all related to patent protection—can create further obstacles to technological innovation and have pervasive social costs.[46]

Among the specialised literature, much effort has gone into the aggregation of patent data and its quantitative study in the long term. Many historians and economists have used patent statistics as a proxy for technological progress, aggregating patent data and drawing international comparisons. Technological capabilities and technological dependence have been some of the issues examined through an analysis of the number of patents granted in various countries.[47] From this perspective, patent counts are a reliable historical source that enables reciprocal long-term comparisons that would otherwise be difficult.[48]

International comparisons of nineteenth-century patent statistics, meanwhile, are problematic. Attempts to quantify innovation present several methodological shortcomings.[49] First, there is no linear relationship between invention and innovation. Not all the inventions devised in the nineteenth century were patented, and among those that were, many were never applied industrially or commercially. Patented inventions were only part of the technologies in use at production sites during the nineteenth century. This is especially true for countries at the periphery of industrial development. Second, the drawbacks and biases of patent statistics may be more pronounced in those societies, such as Spain, that have supported industrial development not in the realm of inventive activity but in the imitation, adaptation and diffusion of foreign technologies. Third, not all countries had comparable patent regulations, and many others did not have any at all or introduced them later in the nineteenth century, so comparisons can be difficult if not impossible (see Table 4.1 in Chapter 4). More important, national patent regulations and institutional arrangements differed significantly and hence produced considerably different outcomes. For example, many countries, such as Spain, did not undertake examinations for novelty and granted patents to importers of technology. That said, comparative patent measurements could indicate long-term general patterns of innovation and serve as a starting point for more nuanced historical studies about the changing interplay among institutions, technology and industrialisation. Otherwise stated, patent quantification is important but cannot be an objective in itself.

# Institutionalising Patents

In one of the introductory paragraphs of *The Making of the English Working Class*, published in 1963, Edward Thompson explains that the word 'making' embraces the idea of 'an active process, which owes as much to agency as to conditioning'.[50] The word 'making' also nicely conveys the process that is the focus of this book: the institutionalisation of patents in modern Spain. The nineteenth-century Spanish system was an institution in the making, shaped and reshaped by the changing historical circumstances and actors involved in its organisation. Given modern Spain's relative technological backwardness, the history of the Spanish patent system should be seen not just as another example of an economic institution devoted to the stimulation of invention but above all as a significant historical case of an institution that was utterly redefined by conditions of backwardness.

This book provides a critical re-examination of the historical evolution of the patent system in modern Spain. It illustrates how the nineteenth-century Spanish patent institution was the product of a particular historical conjuncture in which industrial leaders such as Britain, Germany, France and the United States concentrated the greatest number of patents whilst relatively backward countries such as Spain had peripheral and small systems very open to foreign interventions. This book makes the case that the degree of industrial backwardness in nineteenth-century Spain set the stage for the process of institutionalisation of its patent system between 1826 and 1902.

Instead of presenting the causality as running from institutions to economic performance, this book explores the inverse relationship, that is, how Spain's technological backwardness and dependence set the stage for the institutionalisation of its patent system. From this perspective, institutional evolution derives largely from underlying economic conditions and structures. Of course—given certain economic and technological imperatives—governmental policies and the wider cultural environment likewise affect institutional mechanisms and institutional inertia. International patterns of technological development, industrialisation and trade largely explained the distinctive peripheral character of Spain's patent system and its institutional organisation during the nineteenth century. This institutionalisation process entailed the introduction of a new technological culture, social infrastructure and narrative that supported intellectual property rights.

Since the nineteenth century, technological innovation and patenting have been a fundamentally global dynamic, as recent historiography has shown. Take for instance research by Zorina Khan, Ian Inskter, Susan Sell and Eda Kranakis.[51] This book participates in this endeavour of internationalising the history of patents, in this case, by locating the Spanish experience in its broader international and colonial contexts. The book's concentration on the particular institution of patenting sheds light on the severity of the constraints to industrial innovation in Spain in the late nineteenth century. The Spanish case is particularly compelling because of this country's location in the so-called European periphery and also because of the centrality of its colonial dimension, which I explore particularly in relation to Cuba.

During the last two decades, economic historians Patricio Sáiz González and José María Ortiz-Villajos have provided a very detailed analysis of the quantitative outcome of the Spanish patent system throughout the nineteenth and early twentieth centuries.[52] The central focus of their important studies has been the long-term patterns of patenting in Spain and the relationship between patent activity and economic performance. Saiz's and Ortiz-Villajos's initial major works and subsequent studies have also made crucial contributions to the history of technology in modern Spain. Departing from the premise that all patent systems were institutionally similar, these studies have instead taken a comparative quantitative approach between the Spanish case and other national cases, particularly in Europe and the United States. These studies show that modern Spain was extremely dependent on both foreign organisational and machine technologies in developing its own industry.[53] However, their quantitative and overarching frameworks need to be complemented by detailed socio-cultural and institutional studies of Spain's nineteenth-century patent system and, more broadly, that country's modern technological history.[54]

Building on the important quantitative studies already available, this book attempts, in turn, to assess the socio-cultural and political dimension of the Spanish patent system, showing how this institution was established and how it evolved following international patterns of globalisation and dependency during the nineteenth century, especially its final decades. This book is first and foremost an analysis of the institutional framework within which patenting took place in Spain between 1826 and 1902. This study is concerned less with the evolution of patenting rates than with the ways that the patent system was established, organised and maintained. Likewise, there is more concern here in understanding how social actors and interest groups viewed and debated patent rights in Spain than in describing the patterns of patenting.

Although aggregated data on patents are used throughout the book, most of what follows consists of interpreting and contrasting case studies and qualitative material. Using Spain and its colonies as a case study, this book also shows the profound institutional diversity that prevailed among national patent systems during the nineteenth century. This book additionally looks for interactions among systems and within an international context, such as technological and institutional transfers. This analysis of the international linkages among national patent institutions aims to provide a corrective to purely national studies of patent systems, in line with recent works that have favoured a global approach in the history of technology and intellectual property rights. Of course, any international analysis of a topic that deals with distinct legal traditions needs to be clear about terminology. It is worth noting that in this book the terms 'industrial property' and 'intellectual property' are used indistinctly. 'Intellectual property' is the term used in the British tradition while 'industrial property' is the term used in modern Spain and other continental European countries such as France and Italy. 'Industrial property' was also the term used in international agreements and conferences on this subject during the late nineteenth century, including the Paris Union of 1883. It comprises patents, trademarks, utility models and protection for industrial design but excludes copyrights.[55]

This book consists of six chapters, including this introductory one. Chapter 2 addresses the key institutional features of the Spanish patent system and provides an overview of the socio-political context in which the regulation of intellectual property rights developed. After examining the system's formal legislation and tacit rules, the chapter traces the public political and doctrinal discussions about patent regulation and reform that occurred between 1826 and 1902, particularly during the 1850s and 1860s. The chapter also aims to understand the motivations and assumptions of those who created and shaped the Spanish system to function the way it did during the nineteenth century. The patent system was established and preserved through both pragmatic legal arrangements and the development of a rhetoric to justify patents. The final section pays attention to the evolving patenting activity in the Spanish system, including the nature of its patent culture and the role of patentees in shaping the system. The latter part of this chapter draws on evidence gleaned from thousands of original patent files for the period 1826–1902, all located in the historical archive of the Spanish Patent and Trademark Office.

Chapter 3 examines the social context in which patenting took place, paying particular attention to the sites of innovation, information provisions and social actors that comprised the system. This chapter revolves around several themes that, taken together, illuminate how the Spanish patent system was institutionally constructed and gradually evolving. It shows that although the patent office had limited effectiveness in Spain as an engine for technical innovation, there developed around this institution an informational infrastructure and a new technological culture consisting mainly of engineering consultancies, intermediary agents and specialised mechanical periodicals. Especially noteworthy is the emergence—in the 1880s and 1890s—of a community of patent practitioners with substantial expertise in securing, commercialising and exploiting intellectual property rights in this country. This chapter ultimately sheds light on the nature and timing of the historical process of professionalisation of science and technology in the European periphery.

Chapter 4 yields an understanding of the peripheral place of the Spanish patent institution in the context of the Atlantic world economy of the nineteenth century. The focus of this chapter is both the international dynamics and transnational actors surrounding the Spanish system, such as foreign patenting, technology transfer, multinational companies and transfer agents and agency. The chapter integrates the history of the Spanish patent institution into the broader historical account of the construction of an international patent system, contributing to an understanding of how patent institutions, both in technically advanced and relatively backward nations, became institutionally integrated during the 1880s and 1890s. The international pressures for legal homogenisation during these decades shaped the Spanish system and made it even more open to foreign influences than before. In this chapter, I also focus on several examples of patenting activity and technology transfer in Spain, including those under-taken by the French mining engineer Adrien Chenot, the British industri-alist Henry Bessemer and the various companies set up by the American inventor Thomas Alva Edison. This chapter concludes with a case study of the Elzaburu agency, a Spanish patent business that assisted foreign inven-tors and companies in the late nineteenth century.

Chapter 5 provides an analysis of the regulation, patenting culture and institutional administration of the patent system in colonial Spain. Given that colonial and postcolonial patent dynamics remain largely ignored or at least underexplored in the literature, this chapter highlights the pro-found differences in policy, administration and philosophy that charac-terised patent institutions in colonial and postcolonial nations. The divergence of patent institutions in Cuba, Puerto Rico and the Philippines during the central decades of the nineteenth century was a consequence of colonial economic imperatives as well as locally prevailing colonial interests and ideologies of those most affected by patent regulations such as agrarian elites. This chapter also traces the institutional reorganisation of the colonial patent system in these three territories during the late nineteenth century, which was a period of both institutional convergence of intellectual property rights at the international level and expanding US influence in the Caribbean and the Pacific.

Finally, Chapter 6 provides a conclusion for this book. This final part of the book offers general commentaries stemming from the study of the process of institutionalising patent protection in nineteenth-century Spain. This study of the Spanish patent system, as well as its colonial and international dimensions, may well be of broader significance and enables general assertions about the political economy of intellectual property rights in industrial, relatively backward, colonial and postcolonial nations. This final chapter also takes to task studies that have conducted interna-tional comparisons of patent statistics during the nineteenth century without regard to the fact that national patent regulations, administrative practices and institutional agency differed significantly among countries, with each patent system producing different results. The epilogue con-cludes with a discussion of the often neglected but important normative question of whether patent protection was a necessary technological pol-icy for Spanish industrialisation during the nineteenth century.

# Notes

1. J. A. Auerbach, *The Great Exhibition of 1851: A Nation on Display* (New Haven: Yale University Press, 1999); C. MacLeod, *Heroes of Invention: Technology, Liberalism and British Identity, 1750–1914* (Cambridge: Cambridge University Press, 2007).

2. P. Young, *Globalization and the Great Exhibition* (Basingstoke and New York: Palgrave Macmillan, 2009); I. Inkster, *Science and Technology in History: An Approach to Industrial Development* (Basingstoke: Macmillan Education, 1991): 159; See also C. Dickens, 'The Great Exhibition and the Little One', *Household Words* 3 (1851): 356–60.

3. *Official and Illustrated Catalogue of the Great Exhibition of Works of Industry of All Nations* Vol. III, Foreign States (London: The Authority of The Royal Commission, 1851); R. de la Sagra, *Memoria sobre los objetos estudiados en la Exposición Universal de Londres y fuera de ella* (Madrid: Imprenta del Ministerio de Fomento, 1853).

4. F. Cabana, 'La Participació Catalana en les Exposicions Espanyoles i Universals', in Jordi Maluquer de Motes (ed.), *Tècnics i Tecnologia en el Desenvolupament de la Catalunya Contemporània* (Barcelona: Enciclopèdia Catalana, 2000).

5. J. Yllas Vidal, *Una ojeada a la Exposición Universal verificada en Londres* (Barcelona: Imprenta Hispana, 1852).

6. *La Ilustración*, No. 10 (06/03/1852).

7. *The Illustrated London News* (10/05/1851).

8. K. Ferris, 'Technology, Novelty, and Modernity: Spanish Perceptions of the United States in the Late Nineteenth Century', *Hispanic Research Journal* 11 (1), (2010): 37–47.

9. D. R. Ringrose, *Spain, Europe, and the 'Spanish Miracle', 1700–1900* (Cambridge: Cambridge University Press, 1998). On the image of Spain abroad see S. G. Payne, *Spain: A Unique History* (Madison: The University of Wisconsin Press, 2008). A general criticism of cultural arguments in the history of economic development can be found in H-J. Chang, *Bad Samaritans, The Myth of Free Trade and the Secret History of Capitalism* (London: Bloomsbury Press, 2010).

10. D. Pretel, 'Invención, nacionalismo tecnológico y progreso: el discurso de la propiedad industrial en la España del siglo XIX', *Empiria: Revista de Metodología de las Ciencias Sociales*, 18 (2009): 59–83.

11. R. de Morenés, '¿Debe ser España un País Industrial?', *La Gaceta Industrial*, No.1 (1865).

12. M. Dánvila y Collado, *Propiedad Intelectual* (Madrid: Imprenta de la Correspondencia de España, 1882): 469.

13. For an interesting discussion on Spain's historical lack of indigenous inventive activity see S. López and J. M. Valdaliso (eds.), *¿Qué inventen ellos?: Tecnología, empresa y cambio económico en la España contemporánea* (Madrid: Alianza Editorial, 1997). On the peripheral character of science and technology in the history of Spain and other countries of Southern Europe see K. Gavroglu et al., 'Science and Technology in the European Periphery: Some Historiographical Reflections', *History of Science* 46 (2), (2008): 153–175.

14. L. Prados de la Escosura, *Spanish Economic Growth, 1850–2015* (London: Palgrave Macmillan, 2017).

15. J. Nadal, *El fracaso de la revolución industrial en España, 1814–1913* (Barcelona: Ariel, 1999); N. Sánchez Albornoz, *The Economic Modernization of Spain* (New York: New York University Press, 1991); G. Tortella, *El desarrollo de la España contemporánea: Historia Económica de los Siglos XIX y XX* (Madrid: Alianza Editorial, 2004). For the idea that Spain developed a considerable industry throughout the nineteenth century, see Ringrose (1996), Op. cit.

16. See, for instance, J. Domenech and J. R. Rosés, 'Technology Transfer and the Early Development of the Cotton Textile Industry in Nineteenth Century Spain', in T. Hashino and K. Otsuka (eds.), *Industrial Districts in History and the Developing World* (Singapore: Springer, 2016): 25–41 and Thomson, J. K. J. 'Transferencia tecnológica en la industria algodonera catalana', *Revista de Historia Industrial* 24 (2003): 13–50.

17. J. A. Miranda and B. Montano, 'Technological Innovation in Industrial Districts in Spain during the First Third of the 20th Century', *Revista de Historia Industrial* 66 (2017): 127–157

18. J. Maluquer de Montes, 'Crisis y recuperación económica en la restauración (1882–1913)', in F. Comín et al. (eds.), *Historia económica de España, siglos XIX y XX* (Barcelona: Crítica, 2002): 243–84; J. Pan-Montojo, 'El atraso económico y la regeneración' in J. Pan-Montojo (ed.), *Más se perdió en Cuba: España, 1898 y la crisis de fin de siglo* (Madrid: Alianza Editorial, 2006): 267–340.

19. For the idea of a nineteenth century dichotomised Spain (urban-rural, periphery-interior, traditional-modern, political power-economic power) see, for instance, N. Sánchez-Albornoz, *España hace un siglo: Una economía dual* (Madrid: Alianza Editorial, 1977).

20. S. Riera, 'Industrialization and Technical Education in Spain 1850–1914', in R. Fox and A. Guagnini (eds.), *Education, Technology and Industrial Performance in Europe, 1850–1939* (Cambridge: Cambridge University Press, 1993): 141–70.

21. E. Fernández de Pinedo and R. Uriarte, 'British Technology and Spanish Iron Making During the Nineteenth Century', in C. Evans and G. Rydén, *The Industrial Revolution in Iron* (Aldershot: Ashgate, 2005): 151–172.

22. See, for instance, Domenech and Rosés (2016), Op. cit.; F. Cayón, 'La introducción de la tecnología eléctrica en la España del siglo XIX', *Actas VII Congreso de la Asociación de Historia Económica* (2001); and J. M. Valdaliso, 'Las navieras españolas en el espejo británico (1860–1914)', *TST* 13 (2007): 94–121; J. Martínez Ruiz, 'La mecanización de la agricultura española: de la dependencia exterior a la producción nacional de maquinaria (1862–1932)', *Revista de Historia Industrial* 8 (1995): 46–64

23. G. Tortella, 'La iniciativa empresarial, factor escaso en la España contemporánea', in P. Martín Aceña and F. Comín (ed.), *La empresa en la historia de España* (Madrid: Editorial Civitas, 1996): 49–60; G. Tortella et al., *Educación, instituciones y empresa: los determinantes del espíritu empresarial* (Madrid: Academia Europea de Ciencias y Artes, 2008).

24. Eric Hobsbawm's remarks on this question are eloquent: 'It is often assumed that an economy of private enterprise has an automatic bias towards innovation, but this is not so. It has a bias only towards profit'. Quoted in W. J. Baumol, 'Entrepreneurship, Productive, Unproductive and Destructive', *Journal of Political Economy* 98 (5), (1990): 893.

25. See, for example, D. Acemoglu and J. A. Robinson, *Why Nations Fail?: The Origins of Power, Prosperity, and Poverty* (New York: Crown Business, 2012).

26. M. J. Daunton and F. Trentmann, 'Worlds of Political Economy: Knowledge, Practices and Contestation', in M. J. Daunton and F. Trentmann (eds.), *Worlds of Political Economy: Knowledge and Power in the Nineteenth and Twentieth Centuries* (Basingstoke: Palgrave Macmillan, 2004): 2–18; D. North, *Institutions, Institutional Change and Economic Performance* (Cambridge: Cambridge University Press, 1990); E. Ostrom, *Understanding Institutional Diversity* (Princeton: Princeton University Press, 2005).

27. H-J. Chang, 'Institutions and Economic Development: Theory, Policy and History', *Journal of Institutional Economics* 7 (4), (2011): 473–498.

28. A. Gerschenkron, *Economic Backwardness in Historical Perspective* (Cambridge, MA: Belknap Press of Harvard University Press, 1962).

29. I. Inkster, 'Technology in World History: Cultures of Constraint and Innovation, Emulation, and Technology Transfers', *Comparative Technology Transfer and Society* 5 (2), (2007): 108–127.

30. J. Mokyr, *The Lever of Riches: Technological Creativity and Economic Progress* (New York: Oxford University Press, 1990): 114.

31. M. Artola, *La burguesía revolucionaria, 1808–1874* (Madrid: Alianza Editorial, 1978); P. Sáiz, *Propiedad industrial y revolución liberal* (Madrid: OEPM, 1995).

32. I. Inkster, 'Technology in History: Case Studies and Concepts circa 1700–2000', in R. Narasimha et al. (eds.), *The Dynamics of Technology: Creation and Diffusion of Skills and Knowledge* (New Delhi: Sage Publications, 2003): 21–84.

33. See T. Pinch, 'Technology and Institutions: Living in a Material World', *Theory and Society* 37 (5), (2008): 461–483; and N. Rosenberg, *Perspectives in Technology* (Cambridge: Cambridge University Press, 1976).

34. I. Inkster, 'Patent Agency: Problems and Perspectives', *History of Technology* 31 (2012): 89–97.

35. C. May and S. K. Shell, *Intellectual Property Rights: A Critical History* (London: Lynee Rienner Publishers, 2006); D. Pretel, 'La economía política del sistema español de patentes en perspectiva internacional, 1826–1902', *Investigaciones de Historia Económica* 13 (3), (2017): 190–200.

36. P. A. David, 'Intellectual Property Institutions and the Panda's Thumb: Patents, Copyrights, and Trade Secrets in Economic Theory and History', in M. Wallerstein et al. (eds.), *Global Dimensions of Intellectual Property rights in Science and Technology* (Washington, DC.: Nacional Academy Press, 1993): 19–61.

37. H-J. Chang, 'Intellectual Property Rights and Economic Development: Historical Lessons and Emerging Issues', *Journal of Human Development* 2 (2), (2001): 287–309.

38. J. Robinson, *The Accumulation of Capital* (London: Macmillan, 1969): 87.

39. D. C. North, *Structure and Change in Economic History* (New York: Norton, 1981); D. C. North, 'A Recommendation on How to Intelligently Approach Emerging Problems in Intellectual Property Systems', *Review of Law and Economics* 5 (3), (2009); K. Arrow, 'Economic Welfare and the Allocation of Resources for Invention', in

*The Rate and Direction of Inventive Activity: Economic and Social Factors* (Princeton: NBER, 1962): 609–626.

40. Mokyr (1990), Op. cit.; M. Boldrin, *Against Intellectual Monopoly* (Cambridge: Cambridge University Press, 2008).

41. Chang (2001), Op. cit.

42. The literature on this topic is extensive. See, for example, P. Moser, 'Innovation without Patents: Evidence from World's Fairs', *The Journal of Law and Economics* 55 (1), (2012); A. Nuvolari, 'Collective Invention During the British Industrial Revolution: The Case of the Cornish Pumping Engine', *Cambridge Journal of Economics* 28 (3), (2004): 347–363.

43. See B. Andersen, 'The Neglected Patent Controversies in the Twenty First Century', *Revista Brasileira de Inovaçao* 2 (1), (2003): 35–78; S. J. Patel, 'The Patent System and the Third World', *World Development* 2 (9), (1974): 3–14.

44. D. Harvey, 'The Fetish of Technology: Causes and Consequences', *Macalester International* 13 (7), (2004): 25.

45. V. Shiva, *Patents: Myths and Reality* (New Delhi: Penguin Books, 2001).

46. A. B. Jaffe and J. Lerner, *Innovation and its Discontents* (Princeton: Princeton University Press, 2004).

47. I. Inkster, 'Patents as Indicators of Technological Change and Innovation – An Historical Analysis of the Patent Data 1830–1914', *Transactions of the Newcomen Society* 73 (2), (2004): 179–208

48. J. Streb, 'The Cliometric Study of Innovations', in C. Diebolt and M. Haupert (eds.), *Handbook of Cliometrics* (Berlin: Springer Reference, 2016): 447–468; Z. Griliches, 'Patent Statistics as Economic Indicator', *The Journal of Economic Literature* 28 (4), (1990): 1661–1707.

49. Inkster (1991), Op. cit., 8–14 and Inkster (2004), Op. cit.; For a general criticism of invention-centred historical studies see D. Edgerton, *The Shock of the Old* (Oxford: Oxford University Press, 2007).

50. E. P. Thompson, *The Making of the English Working Class*, Rev. edn. (London: Penguin, 1980): 8.

51. S. Sell, *Private Power, Public Law: The Globalization of Intellectual Property Rights* (Cambridge: Cambridge University Press, 2003); E. Kranakis, 'Patents and Power: European Patent-System Integration in the Context of Globalization', *Technology and Culture* 48 (4), (2007): 689–728; Z. B. Khan, 'Selling Ideas: an International Perspective on Patenting and Markets for Technological Innovations, 1790–1930', *Business History Review* 87 (2013): 39–68; I. Inkster, 'Technology

Transfer in the Great Climacteric. Machinofacture and International Patenting in World Development, circa 1850–1914', *History of Technology* 21 (1999): 87–106.

52. P. Sáiz, *Invención, patentes e innovación en la España contemporánea* (Madrid: OEPM, 1999); J. M. Ortiz-Villajos, *Tecnología y desarrollo económico en la historia contemporánea: Estudio de las patentes registradas en España entre 1882 y 1935* (Madrid: OEPM, 1999).

53. P. Sáiz, 'Los orígenes de la dependencia tecnológica española. Evidencias en el sistema de patentes (1759–1900)', *Economía Industrial* 343 (2002): 83–95; J. M. Ortiz-Villajos, 'Spanish Patenting and Technological Dependency, pre-1936', *History of Technology* 24 (2002): 203–32.

54. For some examples of this approach see D. Pretel, 'El sistema de patentes en las colonias españolas durante el siglo XIX', *América Latina en la Historia Económica,* Vol. 26 (2), (2019); D. Pretel and P. Sáiz, 'Patent Agents in the European Periphery: Spain (1826–1902)', *History of Technology* 31 (2012): 97–114; Pretel (2009 and 2017), Op. cit.

55. For a conceptual history of intellectual property rights and the distinction between patents and copyrights see P. O. Long, 'Invention, Authorship, Intellectual Property and the Origin of Patents: Notes toward a Conceptual History', *Technology and Culture* 32 (4), (1991): 846–884.

# 2

# Making the System

**Abstract** The Spanish patent system was established, developed and maintained through pragmatic legal regulations and public rhetoric that supported intellectual property rights on inventions. This chapter considers the key institutional features of the system—including the tacit rules and deliberate administrative arrangements at work. It also seeks to understand the motivations and assumptions of those who created and shaped the Spanish system to function as it did within a context of political and doctrinal discussions about patent regulation and reform and, more broadly, the mid-nineteenth-century European controversy on the matter. The chapter finishes with a study of the evolving patenting activity in the system between 1826 and 1902, including the nature of patent culture and the role of patentees in modelling the system.

**Keywords** Regulation • Political debate • Controversy • Bureaucracy • Patent culture

The nineteenth century was a volatile time in Spain, with changing political regimes and various legal reforms. In the first part of the nineteenth century, Spain experienced a so-called liberal revolution, which entailed the

© The Author(s) 2018                                                          **27**
D. Pretel, *Institutionalising Patents in Nineteenth-Century Spain*, Palgrave Studies in Economic History, https://doi.org/10.1007/978-3-319-96298-6_2

dismantling of the absolutist regime and a gradual, but conservative and incomplete, reform of the old structures of ownership. Spain's bourgeois reform was limited in its scope, as it kept in place an oligarchic political system and made inadequate provisions for the renewal of ruling class and state structures.[1] Although Spain's turbulent political history during the nineteenth century was not the ideal background for industrialisation, the Spanish state deliberately put forth an institutional framework to surmount the obstacles to industrial development.[2] From the late 1820s, a number of regulations were enacted to further the importation of foreign technologies. For example, in 1828, it was required that all foreign machinery and instruments that could be useful for Spanish industry and agriculture were exempt from taxes, except for the technology imported for the purpose of internal trade, which became subject to a 2% tax.[3] Similarly, the tariff on machinery imports was significantly reduced during the period 1841–1891. For instance, the law of 1841 subjected steam machinery to a tariff as low as to 2%. Later, in 1891, new tariff legislation was introduced that stimulated commercial protectionism and defended domestic machinery makers and national industrial production.[4]

Apart from the trading of technology, several institutional reforms were devised during the nineteenth century to foster industrial modernisation through market intervention. Among state institutions, the setter of property rights on invention was of special importance. The patent system was designed in Spain in 1826 with the objective of increasing technological transfer and the dispersal of technical information. The particular way in which patent policies were implemented in Spain was rooted as much in the intrinsic Spanish culture of political economy and conservative liberalism as in its technologically backward condition. Both national and international determinants shaped the system throughout the nineteenth century. However, what was most distinctive about Spain's patent policy was that it was undertaken as a result of the country's perceived industrial needs and technological weaknesses. Spanish governments conceived the concession of patent rights as a pragmatic institutional formula for inducing technological change. The Spanish system was established as an institution for technological imitation as well as a part of a broader information infrastructure for the diffusion of knowledge and technical expertise.

# Regulating Patents

During the sixteenth, seventeenth and eighteenth centuries 'royal privileges of invention' were seldom granted in Spain. These types of monopolistic protection covered both original technical improvements and importations of foreign machinery, but their concession and conditions were arbitrary.[5] It was not until the first third of the nineteenth century that Spain experienced a rather rapid transition from vaguely defined absolutist royal privileges of invention to more reliable and precise regulations concerning intellectual property rights (in this case patenting). The 1826 patent law, similar to those of other European countries at the time, represented an early institutional reform.[6] The patent regulation was established with the same institutional appearance on paper, or *de jure*, as in other European countries such as France, England and Austria. While the regulation of 1820 had laid the foundations for the Spanish patent system, it was the patent law of 1826 that inaugurated a new era of regulation of inventive activity in the country.[7] The 1826 law of 'privileges of invention', passed by the absolutist King Fernando VII, introduced a patent system to an agrarian and relatively technically backward country in the context of the early industrialisation that was spreading across the countries of continental Europe. Even over the course of various political regimes and severe conflicts, the 1826 law remained basically unaltered for 50 years, as it was not until 1878 that a major reform was accomplished, although the essence of the norm remained the same.

Spain's patent system experienced an institutional inertia throughout the nineteenth century. Numerous political regimes maintained the spirit of the 1826 law at least until the patent regulation of 1902.[8] Throughout the nineteenth century, new patent laws were enacted and different registration bodies were established, which did not entail substantial changes in the ways of granting, managing or advertising intellectual property rights. Until 1887, the Real Conservatorio de Artes y Oficios (Royal Conservatory of Arts), founded in Madrid in 1824, was the institution in charge of patent registration and diffusion. Beyond these duties, the Conservatory was a large ancillary institution that was also responsible for such matters as industrial training, business consultancy, knowledge gathering and technological information dispersal. For instance, to promote the public dissemination of new technologies, the Royal Conservatory of Arts organised six national industrial exhibitions between 1827 and 1850.

From 1887—following the requirements of the agreements of the Paris Convention for the Protection of Industrial Property of 1883—the Conservatory of Arts was replaced by the Dirección Especial de Patentes y Marcas (Special Office for Patents and Trademarks). Its separate status was without continuation due to the lack of economic resources and did not mean the incorporation of specialised personnel; if anything, the opposite was true. It was not until the industrial property law of 1902—more complex, ambitious and far-reaching—that a separate patent office was consolidated. From 1826 until 1886, brief news on patent concessions, transmissions and expirations was made public in *La Gaceta de Madrid* (The Madrid Gazette) and provincial official publications. From 1886, this activity of public diffusion was carried out by the *Boletín Oficial de la Propiedad Industrial* (Official Bulletin of Industrial Property), a governmental periodical exclusively dedicated to intellectual property questions.

The Spanish patent system was modelled after the French system. The French revolutionary norm on industrial property, passed in 1791, acted as a legal model for Spain's 1826 law. There occurred a deliberate transfer of a foreign institution and the appropriation of the legal and intellectual arguments supporting it.[9] This transfer of the French patent code was part of Spain's systematic replication of the Napoleonic state's legal and administrative architecture, accomplished during the first four decades following Spain's independence from France in 1808. Two types of patents were granted, depending on the nature of the technical ingenuity: patents of inventions or introduction. The duration of the patent rights was at the will of the patentee: five, ten or fifteen years for domestic inventions and only five years for inventions imported from abroad. Both patentees who were Spanish citizens and those who were foreigners enjoyed the same rights.

Although the French and Spanish laws were built along the same set of basic principles, each legal system had its own institutional identity and administrative practices. The institutionalisation of the Spanish patent system during its formative years (c.1826–1878) was largely endogenous to the socio-economic circumstances of the time. Direct intervention in the dynamics of industrial innovation was regarded by the various governments as necessary for the welfare of the Spanish nation-state. The implications for the design of the Spanish patent system were manifold: it became a peripheral, pragmatic, neo-mercantilist and dual institution.[10] This is clear when we look at the deliberate nuances introduced in the patent regulations passed between 1826 and 1878, which were aimed at improving the country's relatively backward industrial position. Since its establishment, the peripheral nature of the Spanish patent system was manifested in, among other things, the objective of promoting national industrial progress through openness to many kinds of foreign influences.

Spanish patent laws stimulated technology transfer through a variety of neo-mercantilist caveats, including weak legal security, the compulsory working of patents, patents of introduction, utility models and a lack of technical examinations. These features made the Spanish system one that favoured technological imitation. The Spanish law was pragmatic, serving many aims, some of them conflicting. It tried to guarantee—under certain conditions and within certain limitations—a balance between protecting the rights of inventors on one hand and promoting the emulation of foreign inventions on the other. In other words, it encouraged inventive activity while promoting technology transfer and the diffusion of foreign engineering capacities. It was also a dual system. The law of 1826 was extended to the colonies in 1833, establishing a dual metropolitan-colonial system with separate patent offices and administrative practices.

The bureaucratic practice often strayed from the political objective of stimulating innovation through patents. Lax regulations, legal uncertainties and a poor monitoring of rights encouraged technology transfer but at the same time left room for abuses, such as the use of patent protection to import foreign machinery without competition.[11] This explains why new regulations on patents were promulgated during the late 1840s and early 1850s. Among them, new administrative requirements were established, particularly state control over the actual exploitation of patent rights during the first year in which they were granted. The penalty for non-usage of a patent right was the cancellation of the monopoly. From 1849, independent notarised reports of compulsory manufacture for patented inventions were required by the Royal Conservatory of Arts.[12] A public notary and an engineer had to certify that patents had been put to industrial use. Despite these stricter requirements, the law provided few practical instruments to guarantee the widespread diffusion and successful assimilation of patented technologies into Spanish local industries.

The period between the mid-1860s and 1874 was one of economic crisis and political instability in which the Spanish patent system failed to gain dimension and socio-economic relevance. It was between 1875 and 1902 that the system transformed. These years were characterised by relative political stability, economic modernisation and industrial take-off, including many significant institutional reforms in, among other things, railway construction, commercial activity and intellectual property rights. Especially important was the patent reform of 1878. This patent law further regulated specific aspects of patent protection, although it did not drastically break with the policy direction of previous regulations.

Following French and British regulations, the new law of 1878 stipulated progressive annual quotas that reduced the costs of patenting. It explicitly recognised a two-year priority right for foreign patentees who had already registered their original inventions in other national systems. It also encouraged the wider diffusion of knowledge through the open disclosure of all the information contained in patent files (technical memoranda, drawings and samples) at the Conservatory of Arts. The new law extended the time of the monopoly to 20 years and regulated protection for incremental additions to patents already granted. The 1878 law, like previous laws (and like the 1902 law to come) accepted as patentable subject matters new techniques and practices (including organisational inventions) in industrial, agricultural, chemical and mining realms but did not accept medicine, natural products or scientific theories. All these stipulations of the law of 1878 were clearly aimed at incentivising patenting, especially from abroad.

The Spanish system remained one of registration throughout the nineteenth century. Replicating the British and French systems, Spanish patent laws did not require any technical examination of novelty from the state.[13] Neither there was any formal examination of utility in place, only administrative requirements. The law left to market forces—that is to say industrialists and entrepreneurs—the responsibility to evaluate whether patented inventions were original, useful and practical knowledge. However, the system did have an implicit requirement of utility insofar as it certified the functioning of patents in Spanish territory within a specified time after they were officially granted.[14] The period in which to prove that an invention had been put into industrial practice was one year between 1826 and 1878 and two years between 1878 and 1902.

As soon as the administrative requirements were met and the official fees paid, patent applications became officially registered monopoly rights. This meant that the system was frequently used as means of publicising new devices and products rather than protecting inventions. Manufacturers used patent certificates as additional proof of the quality and reputation of their products when in reality the state had not assessed the technical utility or novelty of the inventions. Moreover, as the Spanish industrial property laws of 1826 and 1878 did not request technical assessments, both foreign and Spanish inventors could obtain patents for inventions that were not original at all.

The 1826 and 1878 laws also stimulated technological emulation through the grating of patents of introduction. From its establishment in 1826, the Spanish system of industrial property awarded this modality of protection to Spanish residents for importing technological improvements into the country without requiring the authorisation of the original inventor. These types of monopoly rights provided imitators with five years of protection as they attempted to introduce foreign inventions not already in use in Spanish industries or agricultural fields.[15] Patents of introduction were a controversial aspect of the Spanish law, to the point that in 1861 a regulation was passed to limit the abuses associated with this type of industrial monopoly.[16]

In Spain, as well as in other peripheral systems such as those of Mexico and Australia, patent rights were less an institution protecting the natural rights of first inventors than a tool of economic and science policy for the stimulation of the national industry through technology transfer.[17] The use of patent rights as a mechanism with which to promote the diffusion of foreign technologies was not, however, exclusive to societies at the economic periphery. No nineteenth-century patent systems strictly followed the doctrine of the natural right of individuals to the property of their original inventions. More industrially advanced countries, such as England and the United States, also had patent regulations that favoured technological imitation. Along these lines, the noted English mathematician and engineer Charles Babbage acknowledged in 1846, in his influential treatise *On the Economy of Machinery and Manufactures*, that patents were an instrument with which to promote not only original invention but also the importation of machinery.[18]

Especially active in promoting improvements in industrial property laws were engineers, whose profession was closely related to the evolution and organisation of the Spanish patent system during the latter part of the nineteenth century. For instance, in 1900, at the request of the Ministerio de Fomento (Minister of Public Works), the industrial engineer Teodoro Merly wrote a patent bill establishing a new register of industrial property directed by engineers. Among other things, it included the regulation of the activities of intermediaries, particularly engineers. The bill was presented in the Spanish Parliament on 30 November 1900 but was not approved.[19] Patent lawyers, commercial intermediaries and agents were likewise influential institutional makers during the late nineteenth century, lobbying for the regulation of the patent system, particularly the institutionalisation of their role. Agents' demands and efforts during the 1880s and 1890s to regulate their activity proved futile. No regulation of any sort was introduced until the 1902 Law on Industrial Property.[20] As the expert lawyer Francisco Elzaburu recognised in a letter to London's Society of Patent Agents, the reform of 1892 represented a 'great progress' for the country, particularly Spanish professionals working on intellectual property questions.[21]

## Debating Patents

Patent rights were during the middle decades of the nineteenth century a source of intense political debate and public controversy at the European level.[22] Open political and social opposition to patent monopolies was common in continental Europe and England between 1850 and 1875. Various collectives asserted the need for radical reform, and even the abolition, of these intellectual property rights. In England, among those who advocated the abolition was the liberal free-market publication *The Economist*, coinciding with the rise of free-trade positions.[23] In contrast to the intense debate that occurred in continental Europe and especially Victorian England during the mid-nineteenth century, in Spain the social and political discussion about patent rights was unquestionably limited.

Patent laws developed in Spain in a socio-political and economic context that differed from those of countries with more intense debates on intellectual property issues. The minor controversy about patent rights in Spain was influenced by this country's peripheral and relatively background position. Because of Spain's relatively slow path to industrialisation during the nineteenth century, the debates about patent regulation that played out in that country are ambiguous and difficult to trace; they rarely occurred in the public sphere and were less visible than in other European cases, such as England, France, Switzerland and Germany. The public positions opposed to the patent laws of 1826, 1878 and 1902 were mostly missing. The absence of objections indicates that the majority of political and economic elites considered patents a tolerable pragmatic policy with which to promote both industrial and agricultural progress. The enthusiastic pro-patent position that could be found among politicians was well represented by the opinions of Manuel Dánvila, a conservative member of parliament who served as chairman and rapporteur of the patent reform of 1878 and authored *La Propiedad Intelectual* (The Intellectual Property) in 1882.[24] For Dánvila, patents concurred 'with the sciences, letters and arts to the moral and material progress of civilisation'.[25]

Although during the mid-nineteenth century the patent institution was not at the centre of Spanish public debate, a range of discussions and anti-patent positions did arise in Spain during the 1850s and 1860s, coinciding with the controversy that swept Europe.[26] The European patent controversy was first extraneous to Spain but was soon echoed in the public sphere of this country. In Spain, the debate revolved around the doctrinal foundations of patent rights, particularly as they pertained to the process of invention. At the core of the public debates was a moral question as to whether an invention constituted an individual's private property, and if so, what type and for how long. An individualistic model of invention connected with notions of natural law (the belief that inventors have a natural right to a monopoly reward for their original inventive activity) was countered by the more collective perspective that saw unfair for society to reward inventors with exclusive patent rights. Several important details of the Spanish patent regulation were likewise a matter of dispute, as some of the incentives preserved a mercantilist touch and were seen as problematic by some liberal economists and politicians.

The most noteworthy discussions took place at professional associations and liberal institutions as well as among renowned professors of political economy. For instance, the debate on patent rights that occurred between 1865 and 1866 at the Real Academia de Ciencias Morales y Políticas (Royal Academy of Moral and Political Sciences) in Madrid, where prominent intellectuals and public figures discussed the advantages and disadvantages of patent rights.[27] Among them was the economist and politician Laureano Figuerola, the politician Antonio de los Ríos Rosas, the economist and historian Manuel Colmeiro and the politician and liberal jurist Alejandro Oliván. The debate showed that patents were a controversial regulation, susceptible to judgments and with less support than literary or artistic property rights.

The discussion at the Academy took place in the same doctrinal grounds as in other European countries in which the reformist doctrines of the French Michel Chevalier (1806–1879) had become popular among prominent free-market economists and politicians.[28] It was a normative valuation of patent rights, especially the question of whether inventions should be regarded as private property. The discrepancies among participants were visible. Two conflicting ideas were confronted: the notion that patents were a natural individual property right versus a collective conceptualisation of innovation where there was no right to a monopolistic use of inventions. The arguments presented at the Academy of Moral and Political Sciences were fundamentally moral, eliding the matter of patent protection as an instrument of economic policy for the promotion of Spanish industry.

The reporting speaker at the debate at the Academia was the liberal politician and economist Luis María Pastor, who had been President of the Sociedad de Economía Política (Political Economy Society) and the Asociación Librecambista (Free Trade Association). Pastor opened the debate by affirming that the question of patents was 'one of the most important in economic science'. Pastor stated in his initial presentation that the Spanish patent legislation produced many harsh inconveniences because it protected not only original invention but also imitations of foreign technologies. He denounced 'the greed of countless mediocrities, who were anxious to acquire a fortune' through obtaining patents for inventions of unclear practical utility. Similarly, the liberal economist Laureano

Figuerola defended patent rights for original inventions, agreeing with Pastor in seeing patents of introduction as unfair and the law protecting those 'taking advantage of original inventors' as morally unsustainable.[29]

In the debate at the Academia, Luis María Pastor also complained about tedious patent administrative procedures and the continuous misunderstandings of patent law and trials.[30] It is clear that patent officials and the government were and would remain suspicious about the increasing influence of patent lawyers and agents in the functioning of the patent system. For example, in 1886 the Minister of Public Works, Montero Ríos, denounced the continuous abuse of the system by intermediaries who had increased eight-fold the cost of applications for industrial property rights.[31] In Ríos's view, the various 'agent intermediaries' were 'taking advantage of the good faith of inventors' and 'discredited the public administration with tedious procedures'. He therefore made the case for the need for additional rules and regulations that would limit agents' central role and rent-seeking activities.

The minor controversy in mid-nineteenth-century Spain over patents was also influenced by the European debate between free traders and protectionists. Patent rights were a difficult compromise between opposing schools of political economy. For the most liberal economists, intellectual property rights were a restriction of commercial and personal freedom.[32] In Spain, the most active patent abolitionists were the free-trade supporters Gabriel Rodríguez, a professor of political economy and civil engineering, and Cipriano Montesinos, a politician and professor of physics at the Royal Conservatory of Arts, the institution that served as a patent office. Rodriguez was one of the few Spanish public figures who openly demonstrated and intensively campaigned against patent rights. Rodríguez had founded—along with the reputed politician, engineer, writer and economist José Echegaray—the liberal publication *El Economista* (The Economist) in 1856, one of the foremost exponents of the abolitionist position at the European level. Rodríguez was a supporter of Michel Chevalier's reformist positions, as evidenced by his articles published in the Parisian magazine *Journal des Économistes* and the Spanish *La Gaceta Economista* (The Economist Gazette), published by the Sociedad Libre de Economía Política (Free Society of Political Economy) of Madrid between 1860 and 1868.[33] As for Montesinos, his criticism of the patent system went unnoticed, although it can be found in the prologue and notes of the Spanish edition of *Principles of Political Economy* written by the English economist John Ramsay McCulloch (1789–1864), which Cipriano Montesinos translated into Spanish along with the jurist and politician Pedro Gómez de la Serna in 1855.[34] McCulloch was, during the central decades of the nineteenth century, one of the leading European proponents of the natural right of inventors to patent rights.[35]

There were also Spanish liberal economists who took a pro-patent stance. A good example is the position of Benigno Carballo Wangüemert (1826–1864), who was a professor of political economy at both the Escuela de Comercio (School of Commerce) and the Real Instituto Industrial (Royal Industrial Institute), both in Madrid and the latter an industrial engineering school closely linked to the world of patents. His extensive, although not very passionate, defence of patent rights can be found in his *Curso de Economía Política* (Course on Political Economy, 1855) and in public speeches such as *Influencias que ejercen los privilegios de invención* (Consequences of Patents of Invention) that he gave in 1857 in Madrid at the Free Society of Political Economy.[36] The latter speech was part of a debate in which other influential Spanish liberal economists—José Echegaray, Gabriel Rodríguez, Joaquín María Sanromá and Félix Bona— were engaged.[37] Preferring to speak of inventors' *priority* rights rather than *property* rights, Carballo Wangüemert maintained that temporary priority rights should be respected and only expropriated for a vital reason.

Critical positions also included Eduardo Saavedra, a professor of applied mechanics at the school of civil engineering in Madrid, who argued that patent monopolies held back industrial progress. In his 1862 review of the *Anuario de Progresos Técnicos de la Industria y de la Agricultura* (Yearbook of Technical Progress in Industry and Agriculture) by politician and industrial engineer José Canalejas, Saavedra reproached Canalejas for not speaking out against patents.[38] Saavedra reasoned that it was not necessary to award any prize or monopoly right to inventors, given that this kind of protection slowed down technological progress and offered legal protection to those who were not the real inventors but 'thieves on more than one occasion of others' ideas'.

It seems that from the early 1870s, Spanish politicians, economists, industrialists and engineers embraced the idea that patent rights—as a sort of 'neo-mercantilist' mechanism—constituted an acceptable policy with which to promote industrialisation, even when these right encouraged the protection and exploitation of technologies emulated from abroad. This in turn allowed Spanish innovators to take advantage of the 'free rider' opportunities stemming from late development and technology transfer. From that point on, the primary argument in support of the patent system became economic in nature: the temporary protection of inventions was seen by Spanish economic and political elites as a sound policy with which to stimulate innovation, imitation and technology transfer. Moral arguments centring on inventors' natural right to private property for the fruits of their labour were no longer contested. This ideological uniformity does not mean that the patent system in late nineteenth-century Spain did not generate any dissatisfaction or controversy, but the debate was no longer connected with even the mildest of controversies in the public sphere.

The institutionalisation of patents in Spain during the 1880s and 1890s was legitimised through the appropriation of foreign ideas and intellectual currents advocating patent protection.[39] Patent systems were, after all, not just a mechanism of transfer and appropriation of foreign technology but also of transfer and appropriation of foreign cultural values. Global technologies went hand in hand with a new globalising culture. This pro-patent ideology was disseminated through new patent regulations, international world fairs, industrial property associations and international agreements, as well as through specialised journals, professional transactions and treatises. An enormous faith in technological progress, along with the need to stimulate Spanish agrarian and industrial wealth, progressively eclipsed any sort of resistance to patent rights during the later part of the nineteenth century.

The ideology associated with industrial modernity had a different impact on Spain than it did on the core industrial countries. It seems that in the Spanish case, frustration and disappointment over Spanish industrialisation during the late nineteenth century radicalised discourse about patent reform to the point that it resembled that of the most extreme patent advocates in continental Europe. These attempts to impose an optimistic interpretation of the role of patents should be seen not just as a lobbying activity by groups with a vested interest in shaping Spain's national and international intellectual property institutions but, more broadly, as part of the faith in industrial progress that characterised the late nineteenth century. However, it is unlikely that the welcoming support for patent protection by engineers, industrialists and lawyers was merely an intellectual position. Clearly, their interests in the patent system played a major role in their support of the Spanish patent institution. Their opinions were self-serving insofar as they amassed a stream of rents from the functioning of the patent system.

From the late 1870s, as pro-patent advocates became dominant throughout Europe, technical literature and political economy publications played a central role in the reception and appropriation of ideas, cultural values and intellectual currents advocating patent protection in Spain. At the forefront of these ideas was a faith in patent protection as an effective way to promote technological innovation and, by extension, economic wealth and industrial progress. This optimistic belief of a link between technological advancement and intellectual property rights, which dominated the late nineteenth century, existed in a context of increasing globalisation, protectionism and industrial competition among nations.[40] The reception of these dominant ideas among Spanish industrial classes speaks to the limitations of broad cultural explanations for Spanish technological backwardness, yet ultimately it tells little about the economic barriers to, and limits of, late industrial development in Spain during the final decades of the nineteenth century.

A harsh assessment of Spain's weak inventive capacity was also common in many of the articles published in the country's growing technical press during the 1880s and 1890s. A widely held assumption was the close relationship between national ingenuity and national progress, a notion that had also been popular in other European countries since the late nineteenth century. Heralding the 'regenerationist' critique that emerged at the turn of the century, these articles concentrated on general flaws of Spanish culture as well as the country's poor material conditions and environmental limitations for technological and industrial progress. American technological modernity, particularly its breakthroughs in communication and electricity, epitomised the new industrial age idealised by Spanish literate classes. Meanwhile, more ambivalent public backlash positions criticising the negative consequences of technology and new industries were also commonplace in Spain, as they were in other European countries such as Britain and Germany.[41]

Spanish industrialists and manufacturers who resorted to patents in this country had fewer doubts about the importance of patent rights. Among capitalists the primary argument supporting the patent system was protectionist. Patent rights were essentially a means to limit competition and protect national industries. The temporary protection of innovation—often through imitation of foreign invention—was seen as a positive policy with which to stimulate national industrialisation. This doctrinal position contrasted sharply with the argument, popular during the 1850s, that patents should not be used to facilitate the importation or imitation of foreign technologies. For example, the Asturian businessman and industrial engineer José Tartiere y Lenegre, a prominent producer of explosives and the founder of the Santa Barbara company, among many other firms, pointed out in 1894 that the five years of protection granted for the manufacture of foreign inventions in Spain was a particularly critical policy for reducing Spain's dependence on strategic techniques and products.[42] Several years before, in 1854, the metallurgist José Vilallonga, director of the Basque company Ybarra Brothers, had declared that, for his firm, patent protection was an indispensable instrument with which to introduce foreign techniques and compete in international markets, and that in facilitating these essential activities it was serving the Spanish nation.[43]

In the same lines as national manufacturers, the influential industrial engineer Gumersindo Vicuña praised, during the 1880s, the role of patents as a tool for Spanish industrialisation.[44] In his articles, Vicuña criticised those critics dismissing the economic utility of patent rights as a policy with which to stimulate industrialisation. However, for Vicuña patent protection should be limited in time, because according to him 'true geniuses and creators...are very rare' and it was not easy to 'point where originality begins' as the 'the work of progress is continuous'. Vicuña even went on to demand that the government assemble industrial juries composed of engineers and manufacturers who would ensure the actual enforcement of the rights of patentees.

Of course, there were more qualified assessments of the implications of patent regulation for Spanish industrialisation and economic development. A good example of a more nuanced argument was that of the prestigious Catalan politician and jurist José Pellá, the most highly regarded patent lawyer of his time. Pellá was the author of *Las Patentes de Invención y los Derechos del Inventor* (Patents of Invention and the Rights of Inventors), published in 1892 and the most significant monograph on Spanish intellectual property rights of the late nineteenth century. In this book, Pellá pointed out that abolitionist positions were effectively opposed to civilised industrial nations and international agreements.[45] He saw patents as the most effective state policy with which to protect and encourage Spanish industrialisation, ahead of commercial policy.[46]

In another book published by Pellá in 1904, in which he commented on the patent law of 1902, he vehemently explained the relationship among patents, protectionism and industrialisation: 'patents are inherently protectionist and were only enacted to protect the national industry'.[47] As for the controversial question of granting patents to people who had invented nothing but had imitated foreign ingenuity, he rather eloquently observed that through this protection many new industries had come to life in Spain.[48] While Pellá was among the most enthusiastic pro-patent advocates, he nonetheless criticised the popular natural rights argument, the idea of the indefinite monopoly of inventors and the frequent negligent uses of the system.

## Patentees and Patenting Culture[49]

During the first 25 years of the system, some 890 patents were filed in Spain's Royal Conservatory of Arts, half of them of five-year monopolies of introduction for inventions conceived abroad. The assignment and commercialisation of patent rights during this formative period of the system were likewise minimal. From the establishment of the system in 1826 through 1850, only 72 royal privileges of invention were assigned in Spain, that is, approximately 8% of the total privileges granted in this country during this period. Most patents never became innovations, and the level of withdrawal of privileges was extremely high due to the relatively high costs of registration and the limited value of patent protection for inventors.

Patent applications were not only administrative procedures but detailed texts on legal, scientific and industrial issues, quite often of considerable length.[50] The patent documentation included a written part, and usually a drawing, whereby the inventor described the new technology and claimed novelty, following the administrative requisites established by the industrial property law of 1826.[51] In principle, the economic value of a patent depended on the skilful writing or translation of patent specifications, since the text was the only basis by which to assess the substantial novelty of an invention in a litigation process. In cases of infringement, the patent specification was the evidence that the new invention was significantly original and did not rely largely on previous knowledge. However, patenting was an expensive activity in Spain as patent rights did not easily become a valuable asset for their holders.

During the formative years of the institution's existence, applications were filed mostly by individual patentees in a language that reflected embeddedness of these individuals in a practical manufacturing culture and their lack of formal technical training. The technical memoranda presented along with patent applications were still based on tacit knowledge and practical experience; explanations were neither formalised nor standardised, reducing the likelihood that third parties could replicate the inventions. Interestingly, in their applications, many patentees still mentioned the role of divine providence as the ultimate source of their inventions. In this context, engineers and professional draughtsmen were not yet considered necessary consultants in the preparation of patent applications. One exception was the activity of the French industrial engineer Constantino Roy, an expert draughtsman, who started assisting patentees as early as in 1853.

Sometimes patentees were neither the inventors nor the real owners of a patent. 'Patent jobbers', who obtained patents on behalf of their clients, were not uncommon in Spain during the mid-nineteenth century; this was also the case in other countries such as England and the United States.[52] Patent jobbers' activities during these formative years were closely related to foreign patenting, mostly French, but also to a lack of reliable intermediation infrastructures to reduce the costs and time of securing rights. Just as patents were granted to imitators, monopoly rights in Spain were not in practice reserved for true inventors. The law and those administrating it did not provide sufficient resources to ascertain the originality of patented inventions. The Madrid-based lawyer Antonio Blanco was one of those performing patent-jobber activities during the 1850s. However, this activity was not professionalised.

During the 1850s and 1860s, Spain saw an increase in patenting activity. New regulations in varied realms—from railways to banking, from land property to trade—and a relative improvement in industrial conditions sparked greater numbers of patent applications. Yet, the number of patents filed in this country paled in comparison to those of industrial leaders. The incentives to patenting in Spain continued to be minimal relative to leading industrial nations. Similarly, despite the increasing technical complexity and economic value of patented inventions, the market for patent rights continued to be extremely narrow during the mid-nineteenth century.

A law in 1849 had introduced formal requirements to demonstrate the compulsory working of patents, including independent technical reports validated by a notary. However, only 25% of the patents granted between 1826 and 1878 and 28% of those awarded between 1878 and 1900 certified compulsory manufacture. The actual manufacture of patented inventions in the Spanish economy was much smaller, as administrative certification did not equate to widespread industrial use or even actual implementation. To satisfy administrative working requirements, patentees tended to resort to already well-established factories, such as the workshop of the metal firm La Maquinista Terrestre y Marítima in Barcelona. Demands for an extension of the period in which to certify the working of a patent were not infrequent. For example, claiming a lack of necessary technical expertise in Spain, the German industrialist Alfred Krupp in 1875 requested that the Conservatory of Arts grant him such an extension for satisfying the compulsory working provisions of a patent on railways wheels.[53]

The shifting nature of technological relationships and the wave of science-based innovations associated with the Second Industrial Revolution had an important impact on the content of patent applications in Spain. Patented technologies evolved during the central decades of the nineteenth century from the diffusion of practical manufacturing technologies to the setting up of large-scale and capital-intensive machinery. Despite these changes in the nature of the activity and the dimensions of the system, both the totality of patent applications and the number of patent applications per capita in the Spanish system remained far below those of the most industrialised countries throughout the late nineteenth century.[54]

By the 1870s the increasing complexity of technologies had changed the content of patents files in Spain. The use of standardised language, technical drawings, mathematical symbols, chemical formulas, principles of physics and mechanical models became commonplace.[55] There was as well a secularisation of the language that patentees used to explain their inventions. Formal patent specifications became an implicit requisite for effectively securing property rights on inventions. However, patentees tended to consciously limit the technical information provided in technical memorandums, presenting the broadest possible claims of novelty even when their patent requests overlapped with patents already granted to previous patentees. Patent specifications were not very detailed and often lacked the relevant technical specifics of new inventions, making them very difficult to operate for third parties. As patentees had no obligation to provide many technical details and there were no examinations of utility, many technological devices often had minimal merits or no practical value at all. For example, dozens of patents were granted for impossible perpetual motion machines. Criticism of these kinds of unworkable patented inventions, unsupported by modern science, was widespread, as evidenced by articles by, among others, the engineer José Echegaray published in the *Revista de Obras Públicas* (Journal of Public Works) and *La Gaceta Industrial* (The Industrial Gazette) during the 1850s and 1860s.[56]

Until the 1880s, there was not yet a large visible community of inventors residing in Spain and actively making use of the country's patent system, as they did not make substantial economic returns from patenting. It is clear that the persistence of a registration procedure for which no state exam was required and the lack of strict enforcement of property rights collectively explain this meagre patenting activity. However, the main factor that explains the relatively late extension of a substantial community of patentees was the weakness of Spanish industrialisation. The limited scope of the Spanish patent system until the 1880s minimised the development of a market for technology. The patent market was limited both in size and the value of property rights for patentees, meaning that assignments and licences were few.[57] After the introduction of the 1878 law, the number of transmissions of patent rights grew considerably, as did the Royal Conservatory of Arts' total revenue collection (from patent fees), which increased by more than eightfold between 1878 and 1886.[58] While these changes are remarkable, the starting point was low indeed.

From the approval of the 1878 patent law, Spain's system gained in dimension and complexity. This regulation is a good example of how legal variations can sometimes rapidly affect institutional design and organisation. The economic and political transformations that occurred from the 1870s in Spain were likewise a factor in increasing the country's propensity for patenting, especially from abroad. Indeed, national and international changes in the institutional environment between 1878 and 1883 fostered valuable international patenting in Spain. The number of patent applications per year, the trade in property rights and the technical and economic value of the inventions protected increased significantly from that moment. The globalisation of trade and investments, the rise of international technology transfer and Spanish political stability paved the way for a gradual transformation in patenting activity during the 1880s and 1890s. Some 31,000 patents were registered between the patent laws of 1878 and 1902. This was a truly remarkable rise, especially when compared with the approximately 5000 patents that had been registered between the patent laws of 1826 and 1878—more than double the time. Even in the period 1878–1902 itself, there was a steady increase in the number of annual patent applications, from 148 in 1875 to 1897 in 1902. Interestingly, although the number of patent applications in Spain increased considerably after 1878, the number of patents issued annually remained lower when compared to not only leading industrial countries but also other Mediterranean countries such as Italy. For example, in the year 1890, 1295 patents were granted in Spain, 25,313 in the United States, 10,646 in Britain, 9009 in France, 4268 in Germany, 1780 in Italy and 116 in Portugal.[59] Yet it should be noted that such quantitative comparisons can be misleading due to the international variations in patent laws and administrative practices.

One of the main features of Spanish patent culture was the prevalence of foreign patentees, confirming the openness and narrowness of Spain's industrial base.[60] In the period 1878–1900, around 60% of patent applications in Spain were from foreigners, the four main countries of origin being France (17.4%), Germany (11.6%), the United Kingdom (10.5%) and the United States (10%). The degree of foreign patenting in Spain reached an average of almost 70% if we include the patents of introduction on foreign technology for which Spanish residents applied. The increasing foreign activity in the system cannot be explained solely by the advance in economic incentives to patenting in Spain. It was an international trend closely linked to institutional reforms introduced after the international agreements on intellectual property rights of the 1870s and 1880s, particularly after the Paris Convention of 1883.[61] Although some latecomers, such as Germany, Italy and Japan, also presented high levels of patent applications by foreigners, they succeeded in developing, during the last decades of the nineteenth century, industrial sectors with a high concentration of domestic patents. In the case of Spain the transition from foreign to domestic capabilities was not attained.

Throughout the nineteenth century there was a correlation between urban culture and patenting activity. The provinces with the highest number of registered patents were Madrid, Barcelona, Bilbao, Cádiz, Málaga and Vizcaya, the urban centres with the highest population density and industries. The urban environment was a centre of invention and technological innovation. It was a social space more receptive to new technologies due to its superior institutional infrastructure supporting innovation and the social proximity that existed among patentees, industrialists, engineers, lawyers and commercial agents. Madrid held the largest number of patent application between 1826 and 1878, and the second largest between 1878 and 1900, after Catalonia. Madrid was the political, administrative and financial capital and centre of the radial transport infrastructure while Barcelona was the leading industrial city. Something similar occurred in England and France, where patenting was concentrated in London and Paris, cities with a large number of law firms, engineering schools and technical institutions and associations.

Throughout the nineteenth century, the vast majority of patentees in Spain were male. Female patentees were a rarity—less than 2% between 1882 and 1897—although sometimes men were granted patents for inventions that in reality were conceived by women, often family members.[62] Identifying the major Spanish and foreign patentees during the nineteenth century requires a qualitative approach to patent applications that concentrates only on the most valuable or useful patented technologies in both economic and socio-cultural terms. Patents lodged at a higher cost, maintained for longer periods and with higher number of assignments can be considered elite patenting activity.[63] From 1878 there was a considerable increase in elite patenting in Spain, although it remained relatively lower than in France and the United Kingdom and substantially lower than in the United States. Elite patents were also those granted to Spanish and foreign inventors with public visibility for their industrial activity and the economic relevance of their 'breakthrough' inventions.[64] Among the most well-known patentees in the Spanish system during the second half of the nineteenth century were the individual inventors Edward W. Serrell, Charles F. Brush, Isaac Peral, Thorsten Nordenfelt, Nicolaus A. Otto, George Westinghouse, Eugène Turpin and Thomas Alva Edison as well as the companies Bell Telephone, Marconi Wireless Telegraph, Portilla & White, Solvay, Vickers and Siemens.

From the 1880s there was an increasing patenting activity from companies, although independent inventors would continue to play a more important role than companies. Patenting by firms in Spain, which in 1880s represented only 9.9% of the number of patent applications, increased to 22% in the period 1900–1909.[65] Companies patenting before the 1880s tended to be small Spanish firms; the operation of sizeable incorporated corporations was still limited. In the period 1890–1902, patents sought by companies rose to over 16%, thanks to greater participation from German, French, American and British firms. In 1878 only 14 companies had applied for patents, versus 390 in 1902. In the period 1878–1902, among the more active firms in the system were the American manufacturing company Thomson Houston Electric Company and its different European subsidiaries, the German family firm Krupp, the French engineering company Schneider & Cie and the German chemical companies BASF and Bayer.

Patents of introduction (or importation) were used by several Spanish companies, serving as an instrument for transmission of, among other technologies, the Bessemer steel converter and Bell's telephone apparatus. Several Spanish companies also obtained a large number of patents, such as the firm La España Industrial, one of the country's largest and most technologically advanced textile companies during the 1880s and 1890s, with more than 2500 workers. Another Spanish company operating in the system during the late nineteenth century was the Asturian firm Santa Bárbara, which specialised in the industrial manufacturing of weapons and artillery, and which also made use of patents of introduction as means to adopt foreign technologies.

An illustrative example is the patent of introduction granted to the journalist and politician José Gasset y Chinchilla in 1902 for a machine to use in the commercial illustrated press. Although Gasset already had an exclusive agreement worth more than 300,000 pesetas with the original inventor, the German company Koening Baüer, Gasset nevertheless filed an application for a Spanish patent of introduction for this printing press.[66] This machine was, according to Gasset, 'the ideal of every journalistic enterprise', but it could not easily be introduced in Spain because of the costs and expenses associated with this type of industry and without the protection of the state.

The previous examples and aggregated data make clear that patents rights were a central vehicle through which foreign innovations were transferred and adapted to Spanish conditions—although often that occurred through imitations not authorised by the original inventors. In 1879 even the popular magazine *Scientific American*, edited by the New York-based patent agency Munn & Co., denounced French inventors and firms for imitating US inventions and patenting them in Spain, in the face of apathy on the part of American inventors[67]:

> We had confidently expected that the new Spanish law on patents would draw the attention of American inventors toward this country, that to-day offers a wide field for every new practical invention, but I am sorry to see that, with the exception of Edison and a few others, the Americans have not yet availed themselves of the easy facility for taking patents for Spain, where new inventions and new industries are now eagerly accepted and adopted. And while the Americans are thus careless as to their own interests, the French take out and negotiate, in Spain, American patents with insignificant variations.

Patent rights—especially foreign lodgments and patents of introduction—were also used as a strategy for limiting commercial and industrial competition. One telling example of this occurred in connection with the public tender announced in 1894 for the construction of a floating dock at the port of Barcelona. Among the proposals were those of two competing Barcelona-based companies: the Arsenal Civil and the Maquinista Terrestre y Marítima. These two machinery manufacturers filed patent rights applications and litigation procedures in an effort to win the bidding for this major civil engineering project.[68] Arsenal Civil in 1892 and Maquinista Terrestre in 1896 obtained five-year patents of introduction for protecting a similar circular floating dock. This invention for repairing large ships out of water had been originally conceived and patented abroad by the British hydraulic engineers John Stanfield and Edwin Clark. Stanfield & Clark also obtained three patents for the same invention in Spain in 1894 and 1895.[69] In 1896, Maquinista Terrestre deemed Arsenal Civil's patent invalid and denounced it in Spanish courts. After some contradictory and controversial orders from the courts and patent office, the Maquinista Terrestre in partnership with Clarke and Stanfield won the bid for the Barcelona port.[70] It seems that the ulterior motive for all these companies was to, simultaneously, secure the construction of a floating dock at the port of Subic in the Spanish colony of the Philippines.

Although often neglected, a paramount dimension of any patent system is its litigation activity over rights infringements. In the Spanish case, not only was patenting activity relatively limited but litigation was insubstantial throughout the nineteenth century. This tendency was consistent with one of the more significant elements of Spanish patent culture: the weak substantive enforcement of regulations. Litigation activity, it must be said, is extremely difficult to document—even for the late nineteenth century— given the dispersed documentation of the few patent trials that took place. Usually patent files kept at the Spanish office in Madrid contained summaries of litigation and infringement, information sometimes also available in patent lawyers' and agents' registers, documentation and publications.[71]

Patent litigation in the Spanish system of the 1880s and 1890s was characterised by contemporary accounts as costly, time-consuming and dominated by a very few expensive lawyers and consulting engineers.[72] This—together with the limited dimensions of the system and the lack of specialised courts for intellectual property issues—may explain why, according to patent lawyer and jurist José Pellá, patent litigation and jurisprudence continued to examine very rudimentary legal questions during the late nineteenth century.[73] From a more pragmatic perspective, the evidence suggests that such a disregard for the effective protection of patent rights had clear advantages. A weak system of registration, like the one encouraged by Spanish governments, state officials, intermediaries and legal courts throughout the nineteenth century, made it less difficult for local industrialists to acquire foreign technologies and reduced the costs of the bureaucratic administration of patents.

# Notes

1. J. Cruz, 'An Ambivalent Revolution. the Public and the Private in the Construction of Liberal Spain', *Social History* 30 (1), (1996): 5–27; I. Burdiel, 'Myths of Failure, Myths of Success: New Perspectives on

Nineteenth Century Liberalism', *Journal of Modern History* 70 (4), (1998): 892–912.

2. M. Artola, *La burguesía revolucionaria, 1808–1874* (Madrid: Alianza Editorial, 1978): 79–80.

3. See Royal Orders of 07/04/1827 and 08/12/1828.

4. E. Fernández de Pinedo and R. Uriarte, 'British Technology and Spanish Iron Making during the 19th Century' in C. Evans and G. Rydén (eds.), *The Industrial Revolution in Iron* (Aldershot: Ashgate, 2005): 151–172.

5. N. García Tapia, *Patentes de invención españolas en el siglo de oro* (Madrid: OEPM, 1994) and P. Sáiz, *Propiedad industrial y revolución liberal: Historia del sistema español de patentes* (Madrid: OEPM, 1995).

6. Royal Decree of 27/03/1826.

7. Royal Decree of 10/10/1820.

8. Industrial Property Law of 16/05/1902.

9. D. Pretel, 'La economía política del sistema español de patentes en perspectiva internacional, 1826–1902', *Investigaciones de Historia Económica* 13 (3), (2017): 190–200; Sáiz (1995), Op. cit.; On the question of ideological transfer see also I. Inkster, 'Politicising the Gerschenkron schema: Technology Transfer, Late Development and the State in Historical Perspective', *Journal of European Economic History* 31 (2002): 45–87.

10. Pretel (2017), Op. cit.

11. For example, see the following contemporary article criticising patent granting procedure in Spain: 'Patentes sin curso', *La Gaceta Industrial* (10/11/1887): 324–26.

12. Royal Order of 11/01/1849.

13. J. Cabanach, 'El examen previo en las patentes de invención', *Industria e Invenciones* (23/05/1908): 197–198.

14. An example of a natural rights argument for supporting patent protection in Spain is 'Privilegios de invención e introducción', *Memorias de la Sociedad Económica de Amigos del País* (Havana: Real Junta de Fomento, 1847): 89–99.

15. P. Sáiz, 'Did Patents of Introduction Encourage Technology Transfer? Long-term Evidence from the Spanish Innovation System', *Cliometrica* 8 (1), (2014): 49–78.

16. Royal Order of 15/10/61.

17. For Mexico see E. Beatty, *Institutions and Investment: The Political Basis of Industrialization in Mexico before 1911* (Stanford: Stanford University Press, 2011). For Australia, see J. Todd, *Colonial Technology: Science and the Transfer of Innovation to Australia* (Cambridge: Cambridge University Press, 1995).

18. C. Babbage, *On the Economy of Machinery and Manufactures* (London: J. Murray, 1846): 359–361.
19. Diario de Sesiones de Cortes, Senado, November 1900, No.10, 114–117.
20. Industrial Property Law of 16/05/1902 (CLE, nueva serie, T. XII). A translation furnished by the Spanish agent Francisco Elzaburu was published in London by the *Journal of the Society of Patent Agents*, Vol. III, Nos. 32 to 36 (1902); Vol IV, Nos. 37 to 40 (1903).
21. Correspondence between Francisco Elzaburu and the Society of Patent Agents reproduced in the *Journal of the Society of Patent Agents*, Vol. III, no. 31 (1902): 109–10.
22. F. Machlup and E. Penrose, 'The Patent Controversy in the Nineteenth Century', *The Journal of Economic History* 10 (1), (1950): 1–29; C. May and S. K. Shell, *Intellectual Property Rights: A Critical History* (London: Lynee Rienner Publishers, 2006).
23. C. MacLeod, 'Concepts of Invention and the Patent Controversy in Victorian Britain', in R. Fox (ed.), *Technological Change: Methods and Themes in the History of Technology* (Amsterdam: Harwood Academic, 1996): 137–153.
24. M. Dánvila, *La propiedad industrial: legislación española y extranjera* (Madrid: Imprenta de la Correspondencia de España, 1882).
25. Cited in Sáiz (1995) Op. cit., 122.
26. See Pretel (2017), Op. cit. and Sáiz (1995), Op. cit.
27. 'Ventajas e inconvenientes de los privilegios de invención, perfección e introducción. Resumen de la discusión que tuvo lugar entre 1865 y 1866 en la Academia de Ciencias Morales y Políticas', *Publicaciones Históricas de la Real Academia de Ciencias Morales y Políticas*, Madrid (2007): 107–113. See also Sáiz (1995), Op. cit., 116–118 and Pretel (2017), Op. cit., 4–5.
28. M. Chevalier, 'Les Brevets d'Inventions dans leur Relations au Principle de la Liberté de Travail et de l'Egalité' (París: Guillaumin, 1878); Machlup and Penrose (1950), Op. cit., 9.
29. 'Ventajas e Inconvenientes', Op. cit., 107–8.
30. 'Ventajas e Inconvenientes', Op. cit., 108.
31. Preamble of Royal Decree of 2/08/1886 (CLE, Vol. CXXXVII). Also, BOPI No.1 (01/09/1886): 3–2.
32. May and Shell (2006), Op. cit., 116–117.
33. See for example G. Rodríguez, 'Discussion sur la Propriété des Inventions'. *Journal des Économistes* t. XXXIV (26/04/1862).

34. J. R. MacCulloch, *Principios de Economía Política* (Madrid: Sanz y Gómez, 1855).

35. Machlup and Penrose (1950), Op. cit.

36. B. Carballo Wangüemert, *Curso de Economía Política* (Madrid: Imprenta de Pedro Romero, 1855).

37. *Revista de Instrucción Pública* IV, Nos. 12, 13 y 15 (25/12/1858 and 8/1/1859).

38. E. Saavedra, 'Anuario de Progresos Técnicos de la Industria y de la Agricultura', *Revista de Obras Públicas* I (9), No. 10 (1862): 109–111.

39. D. Pretel, 'Invención, nacionalismo tecnológico y progreso', *Empiria* 18 (2009): 59–83.

40. Pretel (2009), Op. cit.

41. K. Ferris, 'Technology, Novelty, and Modernity: Spanish Perceptions of the United States in the Late Nineteenth Century', *Hispanic Research Journal* 11 (1), (2010): 37–47.

42. See his argument in the following patent files: AHOEPM, Patent No. 14554 and Patent No. 16801.

43. AHOEPM, Privilege of Invention No. 1212.

44. G. Vicuña, 'La naturaleza de los inventores', *La Semana Industrial* 53 (5/01/1883): 525–526 and G. Vicuña, 'Complemento de la ley de patentes', *La Semana Industrial* 5 (01/01/1886): 7.

45. J. Pellá, *Las patentes de invención y los derechos del inventor* (Barcelona: Industria e Invenciones, 1892).

46. Pellá (1892), Op. cit., 25–30.

47. J. Pellá, *Nuevo tratado de patentes de invención* (Barcelona: Espasa, 1904): 97.

48. Pellá (1904), Op. cit., 97.

49. The contentions of this section rest on evidence gleaned from thousands of original patent files for the period 1826–1902, all located in the AHOEPM. See Notes on Sources at the end of this book.

50. G. Bowker, 'What's in a Patent?', in W. E. Bijker and J. Law (eds.), *Shaping Technology Building Society. Studies in Sociotechnical Change* (Cambridge, MA: MIT Press, 1992): 53–74; M. Biagioli, 'Patent Republic: Representing Inventions, Constructing Rights and Authors', *Social Research* 73 (4), (2006): 1129–1172.

51. Pretel (2009), Op. cit.

52. For the activities of 'patent jobbers' in the English and American cases see Dutton (1984), Op. cit., p. 96, and Z. Khan, *The Democratization of Invention: Patents and Copyrights in American Economic Development, 1790–1920* (Cambridge: Cambridge University Press, 2005): 31.

53. AHOEPM, Privilege of invention No. 5380.
54. P. Sáiz, 'The Spanish Patent System (1770–1907)', *History of Technology* 24 (2002): 45–79; Pretel (2017), Op. cit.
55. For the formalisation and standarisation of patent specifications in mid-nineteenth-century Spanish patents, see R. Amengual, *Bielas y Alabes, evolución histórica de las primeras máquinas térmicas a través de las patentes españolas, 1826–1914* (Madrid: OEPM, 2008); Pretel (2009), op. cit.
56. See for example J. Echegaray, 'Del movimiento continuo', *Revista de Obras Públicas* No.1, t. I (4), (1853): 43–44 and No. 2, t. I (12), (1854): 145–151; P. Amand, 'El movimiento continuo', *La Gaceta Industrial* 12 (1865).
57. P. Sáiz, *Invención, patentes e innovación en la España contemporánea* (Madrid: OEPM, 1999).
58. R. Teijelo, *El Real Conservatorio de Artes, 1824–1887* (Barcelona: unpublished doctoral thesis, UAB, 2011).
59. Y. Plasseraud and F. Savignon, *Paris 1883: Genèse du Droit Unioniste des Brevets* (Paris: Litec, 1983): 226.
60. J. M. Ortiz-Villajos, 'Spanish Patenting and Technological Dependency, pre-1936', *History of Technology* 24 (2002): 203–32; Sáiz (1999 and 2002), Op. cit.; Pretel (2017), Op. cit.
61. I. Inkster, 'Patents as Indicators of Technological Change and Innovation: An historical Analysis of the Patent Data 1830–1914', *Transactions of the Newcomen Society*, 73(2), (2003): 179–208.
62. I. Tejero, *Historia y género: la participación de mujeres en el sistema de patentes español* (unpublished master thesis, UAM, 2008); J.M. Ortiz-Villajos, *Tecnología y desarrollo económico en la historia contemporánea* (Madrid: OEPM, 1999): 156–158.
63. Inkster (2003), Op. cit.
64. For assessing the relevance of patentees, it is useful to resort to the following biographical sources: *Diccionario Biográfico Español* (Real Academia de Historia, 2011; http://dbe.rah.es/), *Oxford Dictionary of National Biography* (Oxford University Press, 2004; http://dbe.rah.es/) and patentees biographies at http://historico.oepm.es/museovirtual.
65. P. Sáiz and D. Pretel, 'Why Did Multinationals Patent in Spain? Several Historical Inquiries', in P-Y. Donzé and S. Nishimura (eds.), *Organizing Global Technology Flows: Institutions, Actors, and Processes* (New York: Routledge, 2013): 39–59.
66. AHOEPM, Patent No. 32819.

67. 'Spain a field for Machinery and Patents', *Scientific American* 40 (13), (29/03/1879).
68. AHOEPM, Patent No. 13491 and 18521.
69. AHOEPM, Patents Nos. 15512, 17610 and 18354.
70. F. Barca and X. Moreno, *El dic flotant i deposant del Port de Barcelona* (Barcelona, Associació d'Enginyers Industrials de Catalunya, 1993); A. Castillo, *La Maquinista Terrestre y Marítima: personaje histórico* (Barcelona: Seix Barral, 1955): 282–3; M. Rodrigo, 'La industria de construcciones mecánicas en Cataluña: El arsenal civil de Barcelona', *Revista de Historia Industrial* 16 (1999): 163–176.
71. Among the latter are specially interesting J. Alcover, 'Infracciones de Patentes', *La Gaceta Industrial* No.9 (10/05/1883); G. Bolibar, 'Derechos del Inventor', *Industria e Invenciones* No. 102 (12/12/1885): 249–50; A. Ungría, "Querellas sobre Patentes", *El Fomento Industrial y Mercantil* No. 350, (20/12/1900).
72. G. Vicuña, 'Complemento de la ley de patentes', *La Semana Industrial* Vol. V (January 1886): 7 and 'Sindicato de inventores', *Industria e Invenciones* No.139 (28/08/1886): 93.
73. Pellá (1892), Op. cit., p. 43.

# 3

# Organising the System

**Abstract** This chapter looks at the broader social context in which patenting took place in Spain during the nineteenth century. The 1870s marked a pivotal moment in the organisation of innovative activity in Spain. From then on, a social infrastructure facilitating patenting developed around the patent office. A variety of agents—such as intermediaries, lawyers and consulting engineers—placed themselves at the centre of the Spanish institution that granted intellectual property rights, shaping it and adapting it to the conditions and requirements of a peripheral European economy. Meanwhile, specialised technical publications, including an array of patent journals, became the broader cultural infrastructure supporting patentees, companies and intermediaries that used the Spanish patent system.

**Keywords** Engineers • Agents • Intermediaries • Technical press • Patent journals

© The Author(s) 2018                                                        **57**
D. Pretel, *Institutionalising Patents in Nineteenth-Century Spain*, Palgrave Studies
in Economic History, https://doi.org/10.1007/978-3-319-96298-6_3

An array of societies, associations and publications outside the patent system formed the broader social infrastructure supporting the development of the Spanish patent institution, understood as a system of information exchange. The growing participation of various social actors assisting patentees brought fundamental changes in the organisational arrangements structuring the patent system in Spain. Lawyers, engineers, consultants, intermediaries, skilled workers, commercial agents, financiers and business partners composed the backbone of the Spanish system in the period 1878–1902. These professionals moulded the system hand in hand with patentees and the state, although the expansion of the system remained constrained by the narrowness of Spain's industrial base.

Cities played a prominent role in nineteenth-century European industrialisation. Urban environments were focal centres for the exchange of ideas and knowledge, as well as suitable social spaces to accommodate commercial life and industrial activities. Social and geographical proximity reduced information costs and favoured the development of a sociotechnical network around the patent institutions. In Spain, Madrid and Barcelona were the urban centres where technical literature, commercial publications, intermediation and professional associations developed. It seems clear that the higher social densities—such as associative activities—in such urban locations promoted the rise of expert networks through and around the slowly expanding patent system. The establishment of modernising institutions—universities, physics cabinets, botanical gardens, industrial schools, mechanical museums, economic societies and scientific academies—brought to bear the emergence of a new technological microculture that became essential for the functioning of the patent system.[1] Several industrial districts in smaller towns in Catalonia, Valencia and the Basque Country would also concentrate sector-specific innovative microcultures during the late nineteenth and early twentieth centuries.[2]

Particularly important was the establishment of engineering schools and trade associations during the second half of the nineteenth century. The growing processional associations of industrialist, traders and businessmen also contributed to the institutional changes that the Spanish patent system would experience in the later nineteenth century, in the context of an explosion in international patenting and technological flows between societies. The new social infrastructure provided a *milieu* ready for the adoption of new industrial technologies and the organisation of a more complex patent institution during the late nineteenth century.

## Engineering the System

The institutionalisation of the Spanish patent system was associated with the rise of engineering capacities in this country, reflecting the changing nature of technological knowledge from around the 1860s. Although the definition of the engineer was not uniform across nations, it is commonly accepted that during the second half of the nineteenth century there was a growing interaction between science and technological innovation that required qualified technical experts for the emerging science-based industries. However, the relationship between formal science and technology was not unidirectional; practical technical knowledge frequently continued to serve as the base for industrial production.[3] In latecomers, with more formalised technological cultures and direct government intervention in industrialisation, engineers became strategic actors of patent institutions designed to respond to the demands of foreign inventors.[4] Engineers also emerged as a response to large civil engineering projects, from railways to harbours. In some cases, engineers were also active agents in technology transfer, adapting industrial and agricultural innovations to Spanish conditions and maintaining and repairing them.

From the mid-nineteenth century, formal professional training became a strategy of the Spanish state.[5] The successful transfer, adoption and diffusion of patented technologies increasingly relied on the presence of specialist engineers with the know-how to operate complex machinery and implement industrial methods in specific industries. However, despite liberal reforms, the level of education—both technical and general—in nineteenth-century Spain remained far of the rest of Western Europe.[6] In the first half of the century, formal technical training had concentrated on civil and mining engineering and skilled workers. Since its establishment in the 1820s, the Royal Conservatory of Arts itself had been involved not only in the granting of patents but also in engineering training and broader technical dissemination.[7]

It was not until the 1850s that industrial engineering schools were created by the state in Madrid, Barcelona, Vergara, Gijón and Seville—

schools that soon would develop specialisations in mechanics and chemistry.[8] These schools provided some practical training in technical drafting and modelling as they applied to patenting activities.[9] An especially relevant school—insofar as it encompassed the institution that granted patents—was the Real Instituto Industrial (Royal Industrial Institute) established in Madrid between 1851 and 1867. This institution not only contained an engineering school but also a commercial school and the Conservatory of Arts. The first director of the Institute was the engineer and politician Joaquín Alfonso Martí, educated at the École Centrale des Arts et des Manufactures in Paris, who had worked for a decade in different European countries. Eduardo Rodríguez and Cipriano Montesinos—also engineers educated at the École Centrale—played a central role in the establishment of the Royal Industrial Institute of Madrid, particularly in defining the curriculum of industrial engineers. The French influence was also visible in the content of the curricula and the mandatory course in French. For example, one of the official manuals used in this school was the *Practical Draughtsman's Book of Industrial Design* (1853), written by the French consulting engineer and patent agent Jacques-Eugène Armengaud—a book that would be translated to English by the British agent William Johnson.[10] One of the first professors of the Industrial Institute was the industrial engineer José Canalejas, who later would become a member of parliament during the Restoration regime.

The overall Spanish technological system was constrained by the dimension and character of industrial engineering education. The number of graduates at the industrial engineering schools remained minimal during the second half of the nineteenth century. In the late 1860s, all the industrial schools closed except for the one in Barcelona, a situation that caused a sharp decline in the number of graduates for several years, until the School in Barcelona expanded during the last quarter of the century.[11] More important, it seems engineers were unprepared for the practical demands of industry. It is not clear that the training provided by those engineering schools was the most appropriate for Spanish industrialisation. By state initiative, Spanish engineering schools promoted formal scientific and theoretical training, but neglected to teach practical, hands-on skills or applied knowledge, leaving graduates unprepared for the demands of industry. Replicating the French model of elitist and centralised engineering education, the Spanish schools were more focused on promoting theoretical and abstract training in the 'pure' sciences, such as physics, than on teaching the concrete applied skills demanded by industries.[12]

A common complaint of mid-nineteenth-century Spanish critics was the existence of a deficit in the technical knowledge of Spanish engineers and industrial foremen. Among others, Ramón de Manjarrés, director of the School of Industrial Engineers of Barcelona between 1868 and 1891, denounced the fact that engineers and technicians required not only theoretical training but also practical skills, following the example of other European countries.[13] Similarly, the industrial engineer José Alcover, very active in the Spanish patent system during the 1860s and 1870s, saw the lack of both technical training and practical industrial knowledge among workers as the primary barrier to widespread mechanisation in Spain. In 1865 Alcover spoke eloquently on this matter: 'The most difficult obstacle to overcome is the lack of industrial education and practice, of which the most immediate consequence is the absolute lack of Spanish technicians who can operate a machine'.[14]

It was during the last third of the century that a growing number of engineering consultancies were established in Barcelona and Madrid. The activity of these consultancies was closely connected with the development of the patent system. Particularly active were industrial engineers, a small elite community with close ties to the state.[15] Industrial engineers assisted both entrepreneurs and firms, domestic as well as foreign, in evaluating, buying, operating and repairing patented technologies. They were the preferred professionals selected by patentees and the state to certify the practical implementation of patented inventions, which had become a legal requirement since 1849.[16] Industrial engineers were likewise frequent expert witnesses in patent trials. Their involvement with the administrative functioning of the system may explain why engineers were some of the most enthusiastic advocates of patent regulations.

The consolidation of the system during the second half of the nineteenth century—together with the increase in international patenting—stimulated a larger market for engineering services in Spain. From the early 1860s, a number of distinguished engineers, machinists and chemists—often themselves owners of patent rights—started working on patent issues, specifically related to mechanical innovations, electrical technology and dyestuffs. Several of these engineers and scientists working on patent issues had been trained in France, Belgium and Britain and frequently doubled as technical assistants and draughtsmen. The new demands of the Spanish patent system during those years provided career opportunities for the relatively small number of professionals residing in Spain who had a technical and scientific background and shared a common expert language with inventors and industrialists. Particularly industrial engineers assisted individuals and firms in drafting patent specifications and responding to enquiries on the patentability of technologies.

Consulting engineers provided not just administrative guidance to patentees but also technical support. Once the patents were granted to their clients, some of these engineers sold the patented machinery and offered assistance to buyers in operating it and resolving technical snags. The increasing number of engineers working as patent consultants in Spain was a trend that has also been identified for the case of Britain, Germany, the United States and France, and it was closely related to the concurrent development of a series of new chemical, electrical and industrial technologies during the so-called Second Industrial Revolution.[17] Some engineers were also important inventors in their own right, either as individual patentees or as full-time employees of firms patenting in Spain. Before 1878, most engineers patenting in Spain had been foreign, particularly French.[18] Several Spanish engineers educated or working at the Conservatory of Arts and the Royal Industrial Institute filed patent applications in Spain during the second half of the nineteenth century. The ranks of Spanish patentee-engineers included the professor of industrial engineering Ramón José Izquierdo, who was editor of the technical journal *España Técnica e Industrial* (Industrial and Technical Spain), and the civil engineer Manuel Maluquer Salvador, editor of the civil engineering publication *Revista de Obras Públicas* (Review of Public Works). Another frequent patentee was the industrial engineer Antonio Montenegro Van-Halen, educated at the Royal Industrial Institute in Madrid. Montenegro obtained over 30 patents in Spain, many of them hydraulic inventions, and even a US patent in 1907 for power transmission coupling devices.[19]

Engineers devoted a significant proportion of their activity to the management of 'elite' inventions.[20] The unprecedented amount of foreign technology that flooded Spain from the 1870s contributed to the mushrooming of engineering consultancies in the country's two large urban centres: Madrid, where the patent office was established, and Barcelona, Spain's leading industrial city. Some of these engineers had the ability to technically engage with inventors and transform their ideas into textual documents that met Spanish legal and bureaucratic requirements for patenting. Take, for example, the case of the engineer Enrique Disdier Crooke—educated at the University of Liège—who provided advanced technological consultancy on the matter of foreign patents from the 1890s. His activity facilitated the transfer to Spain of a number of key iron and metallurgical innovations and products.[21] Disdier had the ability not only of assisting patentees but to advice clients in selecting, adopting and adapting foreign technologies to local conditions.

A pioneering and large consulting engineering business was El Centro Auxiliar de la Industria (The Industry Assistance Centre), created in 1871 in Barcelona by Teodoro Merly. Merly was a well-known industrial engineer and regular contributor to technical and professional journals like the prestigious *Revista de Obra Públicas*. Two experienced Barcelona-based industrial engineers—Ventura Serra, editor between 1881 and 1883 of the patent journal *Gaceta de la Industria y las Invenciones* (Gazette of Industry and Innovations) and director of the International Patent Agency, and Felix Sivilla, director of the Centro Auxiliar Mecánico (Mechanical Assistance Centre)—joined Merly in his technical office in the late 1870s, all of them working as consultants for foreign individuals and firms such as the Rubber Tire Wheel Company. Merly, Sivilla and Serra had not only international clients, but devoted a substantial part of their activity to assisting Spanish companies such as the Sociedad Farmacéutica Española. These expert engineers tended to participate in various related activities and offered their clients a more extensive range of services (testing, installations and exhibitions) that overlapped with those of electrical and mechanical consultancy and technology trade. One of their principal activities was the representation of Spanish and foreign manufacturers through permanent exhibitions of machinery.

Consulting engineers were also active in the process of technical information searching. Some industrial engineers, like Barcelona-based Gerónimo Bolibar, provided clients with summaries and copies of letter patents, specifications, technical memorandums and drawings of patents already granted, thereby confirming ownership, titles and assignments. This service was most likely a principal source of the technical information contained in patents, alongside with patent journals and related literature, since the complete technical specifications of inventions were not published by the official organs of the patent office. *La Gaceta de Madrid* and, from 1886, *El Boletín de la Propiedad Industrial,* published only the titles and very brief abridgments of the patents granted or assigned—not the technical details of the inventions. In some cases, engineers' assistance to patentees and industrialists was complemented by the services of patent lawyers with experience in assessing patentability, international priority rights and infringement of inventions within the country.

## Agents and Agency

During the foundational years of the Spanish patent system (1826–1850), the proportion of individuals acting as representatives, assistants or rights traders was decidedly low.[22] The limited dimension of the system, as well as its weak enforcement of rights, hindered the participation of actors other than inventors. The system was extremely insubstantial and patents did no carry almost any value to inventors. Although there were several patent applications and assignments granted through intermediation, a professional community with significant expertise on intellectual property rights had not yet emerged. Some people acted as occasional representatives and assistants to patentees, but their activities were less professionalised than in other countries.

In contrast with the French, British and French cases there is little evidence of individuals working on patent issues in mid-nineteenth-century Spain. Until the 1850s no one worked as a patent expert, not even as an additional job or a side business. There was only a very heterogeneous collection of intermediaries not yet professionally engaged as agents. Rather than serving as professional patent agents, these individuals were contracted for broad patenting-related activities as general attorneys, international merchants, mechanical drafters, machinists, model builders, consulting engineers and practical scientists—some of whom were regular patentees in their own right.

Only a few foreign diplomats and engineers residing in Spain occasionally stepped outside their main activities to dabble in patent-related services during these first 25 years of the country's system. For example, Ernest Dalwin, the secretary of the Belgian embassy, and Juan Lagoanere, the ambassador of France in Madrid, were sporadic representatives of inventors from their respective countries. Similarly, professionals such as the Parisian Enrique Mambert, the British merchant Charles Green, the Spanish botanist Ramón de la Sagra and the Irish banker Henry O'Shea were also occasional intermediaries on behalf of foreign inventors in their applications to obtain royal privileges of invention in Spain during the 1830s and 1840s.

As Fig. 3.1 shows, the number of patent applications channelled through intermediation grew rapidly from the 1850s. Although this graph shows only patent applications submitted by inventors who had resorted to a representative (not services of consultancy, assignments or litigation), it reflects the general evolution of intermediary activities in the system. During the 1850s and 1860s, there was a steady rise of intermediation activity, although engineers and lawyers assisting inventors were not yet full-time patent professionals and did not recognise themselves as such. Moreover, direct patenting by Spanish inventors and firms, without resorting to representatives, continued to be commonplace. Until the mid-1870s a broad array of profit-seeking individuals smoothly entered the patent system rather than consolidating a new profession of patent experts—although there is little evidence of their early intermediation activities.

The 1870s marked a transition in the organisation of innovative activity in Spain. There can be little doubt that the new division of patenting activity was related to the increasing opportunities for the collection of rents associated with patenting at the national and international levels. By 1878—when a new patent law had been passed—lawyers, engineers and business agents had placed themselves at the centre of the expanding Spanish patent system. From that moment onward, a community of specialised techno-legal experts emerged in the country's largest cities,

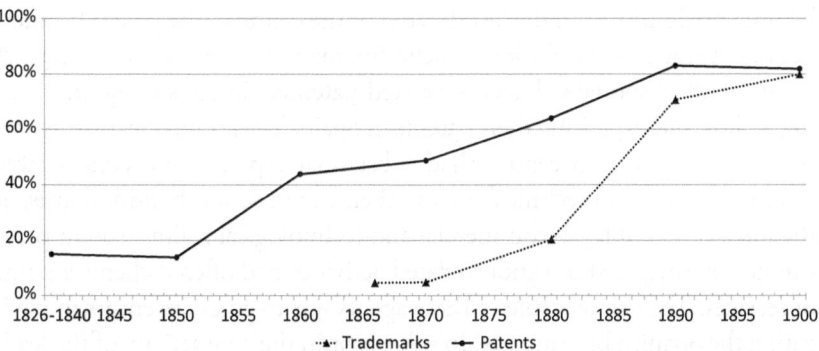

**Fig. 3.1** Patent and trademark applications involving an intermediary, Spain (1826–1900)

becoming critical actors in the functioning of the Spanish patent institution during the final decades of the nineteenth century.

As Fig. 3.1 shows, the percentage of patent applications in Spain involving an agent increased from 50% in the 1870s to over 80% at the century's end. A rising global economy and international agreements on patent rights seem to explain this rise in intermediation during the 1880s and 1890s in Spain as well as other national systems. During these years, patent intermediation also professionalised in other industrial latecomers, including Japan, India, Sweden, Germany and Australia. The degrees of specialisation, training requirements, legal attributions and professional standards required of agents differed across countries, reflecting the international diversity of patent cultures and institutional structures. Sizable professional communities of patent experts were concentrated in just a few large countries, with the United States, France and Britain hosting the overwhelming majority. The number of agents in each national system correlated with their level of patent activity. Early industrial countries, such as Britain, had more informal agencies, whereas late industrialisers, such as Germany, had substantial numbers of formal agencies.[23] As in France and Britain, the Spanish patent agencies focused primarily on patent applications, which were highly bureaucratic, costly and time-consuming as a result of the troublesome nature of Spain's working provisions and the writing of specifications as required by the Spanish law.[24]

Spain's patent business underwent a process of professionalisation between 1878 and 1902. This meant a transition from the generalist patent practitioners of the mid-nineteenth century to the specialised patent agencies that became predominant in the last decades of the century. The patent business became a full-time professional niche for many lawyers and engineers. A community of specialised agents assisted patentees in registering, publicising, selling and exploiting their patents in Spain. The leading Spanish agents of the late nineteenth century had substantial expertise and considerable training in legal and technical issues. Their expertise was hybrid, that is, at the intersection of legal, commercial and technological realms. Patent practitioners participated in various related activities and offered clients a range of services. The primary role of these agents was to guide inventors in navigating the Spanish bureaucracy. In other words, the core activity of the leading patent practitioners was of an administrative nature.

Patent drafting often involved the translation of specifications and the adaptation of drawings, in which case the activities were not merely administrative but substantially contributed to the making of patent rights of invention and the transmission of information from foreign patent institutions to the Spanish one. These patent application-related activities overlapped with others carried out by intermediary experts, including the commercialisation of property rights, the drafting of plans and designs, engineering consultancy, evaluation of inventions, legal advice, technical journalism and assistance in translating patents into functional technologies. From the 1870s a commonality of technical and general press advertisements saw agents selling patent rights and machinery. Agents' co-ownership of patent rights and partnership with inventors were not infrequent. It seems that some of these activities were controversial, with detractors arguing that they conferred excessive power on agents, lawyers and engineers.[25]

Despite the reduced number of trials, there was a small group of patent lawyers concentrating on legal issues and patent disputes.[26] These lawyers assisted patentees in preparing their applications so as to avoid infringement and, eventually, in resolving questions of ownership in legal disputes and trials. They were also among the most respected experts in the interpretation and drafting of Spanish patent legislation. The most renowned patent lawyer at the turn of the century was José Pellá, who provided legal assistance in patent and trademark matters and represented patentees in courts of law in the event of infringement proceedings. Based in Barcelona, Pellá often collaborated with the engineering firm owned by Gerónimo Bolibar.

A look at late nineteenth-century business directories suggests how different the Spanish patent system had become during the last two decades of the nineteenth century. Lists of Spanish patent agents could be located in publications such as the *International Directory of Patent Agents*, published in London between 1893 and 1901.[27] This directory, organised by countries, provided information about agents' locations and in some cases their services, costs, educational backgrounds, past accomplishments and memberships in professional associations in a variety of countries. The expansion of patenting in Spain was also reflected in the rising number of advertisements for engineering consultancies, patent businesses and expert lawyers in the technical, commercial and general press as well as in the official publication *El Boletín de la Propiedad Industrial* (see Fig. 3.2 for an example of these advertisements).

# ANUNCIOS.

**Fig. 3.2** Advertisements for patent businesses, BOPI (1892). (Source: BOPI no.129 (1 January 1892))

Table 3.1 Agencies, agents and consulting engineers, Spain 1860–1900

| | Year founded | City (offices) | Professional background (in origin) |
|---|---|---|---|
| **El Porvenir de la Industria** (*Magín Llados, Federico Cajal, A. Cruset*) | 1857 | Barcelona | Industrial engineers, chemists, politicians |
| **Alfonso Piquet** | 1860s | Madrid | Civil engineer |
| **Elzaburu** (*Julio Vizcarrondo, Francisco Elzaburu*) | 1865 | Madrid | Lawyers, politicians and commercial agents |
| **Clarke, Modet and Co.** (*Alberto Clarke, Fernando Modet, José Gómez-Acebo*) | 1879 | Madrid | Lawyers |
| **El Fomento Industrial y Mercantil** (*Agustín Ungría*) | 1891 | Barcelona Madrid Valencia | Business agent |
| **Oficina Internacional de Propiedad Industrial** (*Gerónimo Bolibar*) | 1884 | Barcelona | Industrial engineer |
| **Centro Auxiliar de la Industria** (*Teodoro Merly, Félix Sivilla, Ventura Serra*) | 1871 | Barcelona Madrid Lisbon | Industrial engineers |
| **Carlos Bonet** | 1886 | Barcelona Madrid | Industrial engineer |
| **José Pellá** | 1880s | Barcelona | Lawyer, jurist and politician |
| **Eladio Pomata** | 1885 | Madrid | Business agent |

   Agents and agencies working on patent issues advertised their qualifications, academic degrees and affiliations with professional associations and institutions. Agents' credentials gave them professional authority and were essential in giving clients a measure of their expertise and proficiency in the field. Remarkably, many patent professionals were prestigious university professors or politicians (see Table 3.1). This may explain why the services provided by these experts were high priced, limiting the number of inventors that could afford their fees.

Over 30 Spanish individuals and firms were consistently listed and advertised in national and international patent directories in the 1890s. However, while business and city directories give a sense of the increasing importance of the industrial property business in Spain, they do not reflect the fact that by 1900 a mere handful of agencies had come to monopolise the patent business via mergers among complementary firms. Lawyers, engineers and commercial agents joined forces during the first decades of the twentieth century, although several engineering firms continued to occupy a separate niche in the market for patent professionals.

The services provided by these agencies frequently differed, as did the professional qualifications and backgrounds of their employees (see Table 3.1). Most agents were engineers, business agents or lawyers, revealing that the changing organisation of patenting went hand in hand with the growth of professionalism and formal education. The overwhelming majority of Spanish agents and agencies during the last decades of the century were practising in Madrid or Barcelona for both economic and administrative reasons.

At the turn of the century, the country's patent agency business was already consolidated and controlled by a few large professional agencies specialising in different types of services. According to patent documentation for the year 1900, a half-dozen industrial property firms controlled about 70% of Spain's total patent applications along with the commercialisation of patents—although the exact percentage is difficult to provide as inventors frequently used chains of intermediary agents and subagents.[28] The leaders of this sector were Clarke, Modet & Co., Elzaburu, and The International Patent Office of Gerónimo Bolibar. It seems that the vast majority of valuable foreign patents were channelled through these leading agencies and engineering firms.

Professional agents also monopolised the process of securing trademarks in Spain in the 1880s and 1890s. From the 1875 Restoration onward, agents' intermediation grew exponentially—to such an extent that at the beginning of the twentieth century they managed virtually every trademark ever registered (93–94%). Four agencies dominated trademark application and management: Clarke, Modet & Co., Elzaburu, Ungría and, mainly, the industrial engineer Carlos Bonet. The latter, based in Barcelona, specialised almost exclusively in applications for, and management of, trademarks. Bonnet channelled approximately 44% trademark applications between 1885 and 1905.

## Institutionalising Agency

What it meant to be a patent agent in late nineteenth-century Spain is unclear. It seems that from the 1880s a sense of belonging to a professional group developed among patent practitioners. However, even during the 1880s and 1890s, many Spanish lawyers and engineers working as patent agents were still not yet professionally engaged as full-time patent professionals and did not recognise themselves as such. As had already occurred in many other European countries, Spanish patent professionals of the 1880s began to use the expression 'patent agent' (*agente de patentes*) in referring to their profession. Agents began to build a professional identity, albeit one built upon a broader range of activities than the narrower one associated with professional patent agents in late nineteenth-century France, Britain and the United States. The concept and professional jurisdiction of 'patent agent' was borrowed from other countries by Spanish collectives working in patent-related activities without necessarily embracing the more precise characteristics or the status associated with this professional category in other countries.[29]

Spanish agents also functioned as a pressure group that eventually became organised into an association. Engineers, lawyers and business brokers attempted to influence reforms of national and international patent regulation through public debate. The construction of patent agents' professional identity in the 1880s was followed by demands—published in specialised journals as well as the mainstream press—for the institutionalisation of their role.[30] The regulation of precisely what responsibilities constituted a patent agent's practice was continually demanded by Spanish agencies during the late nineteenth century. For instance, the engineer Geronimo Bolibar repeatedly called on the Spanish government to 'urgently regulate the class of patent agents instead of trying to eradicate it', since eradication seemed impossible.[31] This engineer, like many other experts with an interest in the patent business, believed that agents were vital for the patent system to function efficiently. Patent practitioners also denounced the government's negative portrayal of agents' activities, along with its legislative passivity.[32] Agents and consultant engineers would demand during the 1890s that the government limit patent intermediation to individuals with formal legal and technical education. They also campaigned for the establishment of state accreditation and a professional association for agents.

Regulation of the patent profession came relatively late to Spain. Neither the patent laws of 1826 and 1878 nor the minor regulations passed in the last two decades of the century had regulated in detail the participation of third parties in the acquisition and management of patent rights of invention. Apart from the acknowledgement that representatives could apply for patent rights in the name of inventors, no other related references can be found in the 1826 and 1878 law. These laws reflected a lack of awareness among legislators of the possibility that actors other than patentees and state officers could participate in the day-to-day activities of the system. The field of patent professionals remained unregulated until as late as 1902. It was with the major reform of the patent law of 1902 that Spanish patent agents' activities became regulated by the state through the establishment of a mandatory register of industrial property agents as well as formal requirements pertaining to their practice. Britain, France, Italy and the United States, among others, had either already adopted some sort of registration procedure for patent agents or at least consolidated a professional association of patent practitioners. Among them, Britain was the first country to introduce a regulatory regime in 1888. Other countries, such as France, did not succeed in imposing an official registration procedure but had a large professional association—although some important agents, such as the lawyer Émile Bert, did not belong to it. This earlier professionalisation of patent agents throughout Europe is not surprising given that these countries had a much more institutionalised and organised inventive community than Spain.[33]

From the 1902 law, the contours of the agents' occupation became both more controlled and more normalised. The 1903 by-law regulating the 1902 Industrial Property Law established both a mandatory register of patent agents and formal requirements for their practice.[34] This regulation established that no individual could serve as an intermediary in more than three files of patent and trademark applications without being registered in the *Negociado Especial de Patentes* (the Spanish Patent Office of the time) as an industrial property agent. The law, however, did not introduce any clear barriers to entry in the profession. It simply specified that only Spanish citizens with a legal or engineering degree or an accredited equivalent competence could be registered as Industrial Property Agents. Registered agents also had to be members of an official Associations of Business Agents (Colegios de Agentes de Negocios) and have a minimum experience of five years working as agents without any judiciary reclamations for malpractice. Nor could they be actual or former civil servants of the patent office.

The 1902 law opened up the patent business to the three types of professionals who had brought the most individuals into the intellectual property field: engineers, business agents and lawyers. From the late 1870s, members of these three occupations had seen a professional opportunity in the patent business. The 1903 patent regulation did not privilege any of them. Engineers managed to preserve the exclusivity of drafting and signing technical memorandums, technical drawings and models that was already regulated by the budget law of 1893–1897, even though the Spanish Patent Office had not acknowledged it.[35] Business agents, engineers and lawyers celebrated the new regulations insofar as they limited patent intermediation to professionals with experience and sufficient economic resources for a mandatory deposit, thereby ensuring a minimum competence in the profession.[36] However, the new regulation left room for abuses. The main issue, according to agents, was that it remained unclear which kind of expertise and qualifications were required of prospective patent agents. The profession of 'industrial property agent', in short, was still fairly vague. In contrast to the precise requirements in Britain, France and the United States, in Spain no qualifying exam or specific training was required.

The register of agents was neither used nor published in the official organ of the Patent Office (*The Bulletin of Industrial Property*) until 1905. This apathy had been denounced by professional agents. A Royal Order in 1905 urged the introduction of a Register of Industrial Property Agents. The remarks of this order were eloquent, arguing that 'the actual situation must no longer continue because it only protects those who [work as industrial property agents] without having fulfilled any sort of requirement', and 'to the detriment of those who from the beginning have fulfilled the legal requirements, thereby offering a guarantee to those who may need their services'.[37]

In the first Register of Industrial Property Agents, released in November 1905,[38] only 28 individuals were registered—21 in Madrid, 6 in Barcelona and 1 in Bilbao—most of them working for one of Spain's five or six largest professional agencies, which had monopolised the country's industrial property business since the last years of the nineteenth century. During the few years following the first registration, the number of individuals registered remained low, with most of them working for only a few patent firms.[39] The first registered agent was Agustín Ungría, who had founded the agency *El Fomento Industrial y Mercantil* in Valencia in 1891. Interestingly, the regulation of agents' activities through the introduction of a mandatory register of agents seems to have stimulated large agencies to absorb smaller ones and increasingly to integrate individuals with technical and legal backgrounds. The regularisation of agent activities also sparked the establishment of new professional agencies, such as Roeb & Co. (which exists to this day), funded in 1904 in Madrid by the agents Leocadio López and Guillermo Roeb.[40]

In a different vein, it was also in the last decades of the century (and the first years of the twentieth century) that Spanish business agents managed to institutionalise their role as recognised professionals in associations and state committees. An early pristine attempt to indirectly institutionalise the patent profession had occurred in 1847 with the creation of the Colegio de Agentes de Negocios (Official Association of Business Agents) in Madrid.[41] In the second half of the century, business agents constituted successive associations that required monetary deposits from members desiring to become full-fledged agents.[42] Between 1895 and 1905 many officially registered business agents, including Pedro Soles Mora, Eladio Pomata, Agustín Ungría, Wanceslao Logares and Enrique Palacio, appeared in the business and city directories of industrial property agents. With these measures, business agents tried to monopolise all kinds of activities related to administrative intermediation with Spanish state offices, including the granting all types of intellectual property rights with a distinctive rent-seeking approach.

It was as late as 1907 that Spanish industrial property agents founded their first professional organisation, La Asociación Española de Agentes de la Propiedad Industrial y Comercial (The Association of Industrial and Business Agents), through the initiative of the lawyer Francisco Elzaburu, the organisation's first president.[43] In May 1909 this agent's organisation acquired official character insofar as it was recognised by the state.[44] It was a very late professional association considering that Britain, France and the United States had already established professional associations by the 1880s, while countries like Australia, Germany, Switzerland and Italy had done so by 1900. The Spanish Association published annual professional reports and its essential task was to collaborate with the state in the reform of the patent system.[45] In December 1916 the Consejo de la Propiedad Industrial y Comercial (Council of Industrial and Business Property) was created as a state consultative cabinet for the reform of the industrial property laws and related issues such as legal reforms and international harmonisation. In 1917, two of Spain's leading patent professionals, Francisco Elzaburu and Agustín Ungría, were among the cabinet's first members, alongside representatives whose ranks included industrial engineers, members of the Academy of Sciences and industrialists. The cabinet was presided over by the former Minister of Public Works, the journalist and lawyer Rafael Gasset.[46] In any case, before their formal institutionalisation—through the establishment of associations and a state council—patent experts had never conformed any kind of professional collective, informal or otherwise.

## Patent Publications and Technical Press

In 1854, the weekly *Scientific American* pointed out that the 'The means which are at the command of inventors at the present day – such as the press – to disseminate a correct knowledge of their improvements throughout the civilised globe, are such as no previous age in the world's history could boast of'.[47] This mechanics' magazine, edited by the leading

American patent agency Munn & Co., is an excellent illustration of technical journalism connected with the global expansion of patent rights during the nineteenth century. Other examples of widely circulated nineteenth-century periodicals devoted to the world of invention are the *Patent Journal and Inventors Magazine*, edited by the London patent business of Barlow, Payne and Parker, the *Practical Mechanics' Journal*, edited by the agent William Johnson and the monthly *Le Génie Industrielle* published by the French ingénieur-conseils from the Armengaud family. These journalistic endeavours served multiple additional purposes: advertising, selling, service offers and a forum for the discussion of patent law. Engineers, patentees, lawyers and agents also published manuals, pamphlets and doctrinal treatises on how to invent and patent, national and international patent laws, practical mechanics and draught, and other issues of interest for the development of the patent systems and the advancement of intellectual property rights. Moreover, the growing number of dictionaries with technological and scientific contents served to communicate knowledge and promote learning.

In Spain, the number and diffusion of publications with a scientific, industrial and technical content experienced a considerable surge in the second half of the nineteenth century, mainly in Madrid and Barcelona.[48] Mechanics and engineering journals and an array of other specialised press became the broader cultural infrastructure supporting the Spanish patent system. From the 1860s, both Spanish publications and foreign ones received in Spain served as a mechanism of knowledge transfer and information spreading. Mechanics' and industrial journals, treatises and pamphlets became nodal points in dispersing practical knowledge among Spanish industrial classes and technical experts.[49] The diffusion of the knowledge present in patent specifications—often generated in other nations—was diffused through these publications, which played a major role in reproducing patent indexes, specifications, abridgements and drawings. The rationale behind the establishment of patent regulations was to disclose the information contained in patent specifications to entrepreneurs and the general population. This was, in effect, the bargain struck between the patentees and society. Patent specifications were thus a significant source of technological knowledge that reduced the costs of collecting information. In this context, technical publications, alongside the patents themselves, served as a complementary means of structuring the technical information of inventions.

Industrial and engineering journals complemented the effort of the official organs—*La Gaceta de Madrid* and, after 1886, the *Boletín de la Propiedad Industrial* (BOPI)—in the dissemination of the technical information contained in patents and trademarks. These periodical publications also commissioned articles on national and international patent laws and other issues related to the world of intellectual property rights. It is difficult to know the extent to which the information contained in Spanish patents was actually important for the development of specific technologies in Spain and its industrialisation process. Yet it is clear that specialised journals served to digest patenting activity and report new technological breakthroughs in a simplified way. These publications were a main channel for the public diffusion of technical information embedded in patent texts along with advertising, training, exhibitions and public lectures.

The centrality of these publications in conveying information on patenting activity can be attributed to the fact that in Spain, in contrast to Britain and the United States, patent specifications and abridgements were not readily accessible in public libraries apart from the one at the Royal Conservatory of Arts in Madrid. Similarly, the increasing number of official exchanges of publications, reports, statistics and manuals between the Spanish patent office and foreign ones was, alongside the catalogues and chronicles of world industrial exhibitions, another way to communicate administrative, legal and technological information to Spain in the late nineteenth century.

The Spanish patent system served other purposes beyond conferring property rights to inventors. Many individual inventors and firms used patent rights more as a means of publicity rather than of protecting technical novelties. Technical journals thus became a central channel of advertisement and trade of machinery in Spain. Patent publications—often in collaboration with patent agencies, foreign firms and commercial houses—published lists of machinery for sale, becoming an efficient vehicle for marketing patented technologies. Technical journalism also provided reliable information about national and international market opportunities for patent rights, such as licencing and assignments, thereby facilitating external investment and partnerships. The public diffusion of technical information and the advertisement of patented technology in journals usually followed the patent application. The invention could not be anticipated in specialised publications without a potential forfeiture of patent rights, as any prior publication could invalidate the patent. Agents from different countries therefore coordinated to secure patent rights in various national jurisdictions before widely publishing the content of the invention in specialised journals.

Several publications devoted largely to the world of invention and patents were published in Spain in the last third of the nineteenth century. The expansion of these specialised presses went hand in hand with the steady stream of international enquiries for international patenting during the late century, as noted in 1885 by the consulting industrial engineer Gabriel Gironi, editor of the publication *La Semana Industrial* (The Industrial Week).[50] Among the Spanish periodicals diffusing patent information, *La Gaceta Industrial* (The Industrial Gazette) deserves particular attention. It was edited in Madrid between 1865 and 1891 by the Catalan industrial engineer José Alcover, author of monographs on technological issues and world fairs, including *La Máquina Moderna* (The Modern Machine), published in 1882.[51] The weekly *La Gaceta Industrial* contained detailed lists of patents, articles on industrial and agricultural methods, and advertisements for foreign machinery and tools available for sale.[52] This magazine was connected with the consulting and machinery trading agency that Alcover oversaw in Madrid, which integrated engineering, commercial and patenting services. An important part of his activity was the commercial representation of foreign mechanical and engineering firms such as Siemens Brothers, The Tanite Company, John Fowler, and Ruston, Proctor & Co.

Almost two decades later, in 1884, another publication closely related to the Spanish patent system was established: the illustrated weekly review *Industria e Invenciones* (Industry and Inventions), edited until 1915 in Barcelona by the industrial engineer and patent agent Gerónimo Bolibar. This publication had an extensive section on the legal and administrative side of the world of patents. It offered summaries of patent activity, technical reports on inventions and comprehensive information on patenting procedures. An earlier example of a publication edited by an occasional patent broker and agent was *La Gaceta de los Caminos de Hierro* (Railways Gazette), whose editor-in-chief during the 1850s was Gustavo Hubbard, an economist with important commercial interests in the development of the railway in northwestern Spain.

Spanish consulting engineers published several mechanical and trade journals intimately connected with the activity of the patent system. Good examples of this trend were the industrial engineer Magín Lladó and the chemist Federico Cajal, both editors of the weekly publication *El Porvenir de la Industria* (The Industrial Future), who in 1857 had established a technical agency in Barcelona dedicated to mechanical and chemical consultancy, valuation of inventions and management of patents and trademarks.

In the 1890s the firm El Fomento Industrial, established by the registered business agent Agustín Ungría, became an important player in the Spanish patent business. El Fomento Industrial was a general commission house not specialising in industrial property rights that saw an opportunity to engage in the commercialisation of patents and trademarks through its commercial journal *El Fomento Industrial y Mercantil*.[53] This publication was active in demanding the reform of the patent law during the 1890s, particularly the introduction of examinations of originality, a longer period in which to prove the viability of a patent and resources to secure the legal enforcement of rights.[54] Working in conjunction with its agency activity, the publication *El Fomento Industrial y Mercantil* had a section where potential patentees could submit queries on intellectual property legislation or practical legal issues.

The study of the audience to whom this specialised press was addressed introduces further complexities. Technical, professional and commercial publications were mostly directed at industrial manufacturers, technical experts and machinery traders. Some publications, however, had a wider audience. Pocketbooks, pamphlets and practical treatises such as the collection on popular mechanics, metalwork and chemistry published by *La Gaceta Industrial* in the 1870s and 1880s found a broad audience among the urban middle classes.[55] Some of these publications served to popularise the inventive and patenting activities of laypeople who were potential independent inventors and by extension buyers and users of technology. A range of works popularised technology, patenting and invention in Spain. Good examples are the collection on popular industry, arts and trade of *La Gaceta Industrial* and Gerónimo Bolibar's practical manuals published in Barcelona in the first decade of the twentieth century.[56]

Others such as the *Boletín de la Asociación Central de Ingenieros Industriales* (The Bulletin of the Association of Industrial Engineers), published from the 1880s, were addressed to industrial engineers and discussed issues of interest for the development of their profession.

Publications with more frequent doctrinal and legislative debates were aimed at the political and economic classes in an attempt to influence the legal regulation and administrative organisation of the industrial property system. Generalist industrial publications exposed the range of views on the patent system and laid out reliable information on foreign patent laws and international agreements. They also had editorial forums in which prominent engineers, inventors, economists and lawyers poured their opinions on a range of controversial issues of patent regulation, such as priority rights, compulsory manufacture of inventions and legal enforcement. For instance, *La Semana Industrial*, set up by the industrial engineer Gumersindo Vicuñar, professor of physical mathematics at the Central University in Madrid, epitomised the role of the industrial periodicals of the day as a leading space for debating industrial and trade policy, including patent regulation. This publication was apparently also important in the reception and explanation of fundamental engineering principles among the Spanish industrial classes.[57]

Some booklets and volumes on international patent laws, very common during the two last decades of the century, were addressed to intermediary experts, thereby facilitating the transnational coordination of patenting activities.[58] These publications offered reliable information about market opportunities for patent rights in different countries. A good illustration of this type of publication was the *Handbook of Patent Law of All Countries*, published from 1884 by the qualified British patent agent W. P. Thomson, which showed the international diversity among patent cultures.[59] Similarly, the publications of agents' associations, such as the *Transactions of the Chartered Institute of Patent Agents* and the *Bulletin du Syndicat des Ingenieurs-Conseils en Matiére de Proprieté Industrielle*, were addressed to practising agents and were a manifestation of their activities as a growing professional group. Several Spanish agents were members of these foreign professional associations. Translations of articles on patent issues and reviews of foreign treatises appeared frequently in Spanish patent journals. For example, in a 1884 edition of *La*

*Gaceta Industrial*, the engineer José Alcover reproduced and commented on an article by the French engineer and patent agent Charles Thirion about the International Convention for the Protection of Industrial Property, held in Paris in 1883, which Thirion had published in *Le Génie Industrielle* a few days before.[60] Spanish journals also advertised foreign mechanical publications such as the *Scientific American*, *The Engineer* and *La Genie Civil*, facilitating their circulation in Spain.

Finally, patent journals were an important mechanism by which to advance the interests of inventors, agents and patentees. A particularly acute and exalting defence of patent rights appeared in an article published in 1885 by the engineer Gerónimo Bolibar, a renowned professional who went on to serve as the secretary of the International Engineering Conference held in Barcelona in 1888. Published under the title of 'The Rights of the Inventor' in the magazine *Industria e Invenciones*, it denounced the state of 'anxiety' undergone by patentees whose rights were not being respected in Spain and called for a true enforcement of the law.[61] In this article, Bolibar argued that patents were not a privilege but a legitimate individual property, analogous to any other social and political rights, such that their usurpation was tantamount to robbery. He also regretted that patent protection in Spain was temporary, costly and penalised non-usage.

**Acknowledgements** Certain parts of this chapter appeared previously as an article in a 2012 edition of *History of Technology* (Vol. 31). This material, which initially had appeared in my doctoral thesis, has been substantially revised for the present chapter. I am grateful to Bloomsbury Academic (an imprint of Bloomsbury Publishing plc) for the permission to reintroduce it here.

# Notes

1. See the contributions to the following volume A. Lafuente et al. (eds.), *Maquinismo ibérico* (Madrid: Doce Calles, 2007) and the following general overview J. M. Sánchez Ron, *Cincel, martillo y piedra: historia de la ciencia en España* (Madrid: Taurus, 1999).
2. J. A. Miranda and B. Montano, 'Technological Innovation in Industrial Districts in Spain during the first third of the 20th Century', *Revista de Historia Industrial* 66 (2017): 127–157.

3. C. Freeman and F. Louçã, *As Time Goes By: From the Industrial Revolutions to the Information Revolution* (Oxford: Oxford University Press, 2001).
4. I. Inkster, 'Engineers as Patentees and the Cultures of Invention 1830–1914 and Beyond: The Evidence from the Patent Data', *Quaderns d'Historia de l'Engyneria* 6 (2004): 25–50.
5. M. Silva (ed.), *Técnica e ingeniería en España* (Zaragoza: Prensas Universitarias de Zaragoza, 2004–2013): Vol. IV–VII.
6. S. Riera, 'Industrialization and Technical Education in Spain 1850–1914', in R. Fox and A. Guagnini (eds.), *Education, Technology and Industrial Performance in Europe, 1850–1939* (Cambridge: Cambridge University Press, 1993): 141–70; J. M. Cano Pavón, *Estado, enseñanza industrial y capital humano en la España isabelina, 1833–1868* (Málaga: Imprenta Montes, 2001); K. H. O'Rourke and J. G. Williamson, 'Around the European Periphery 1870–1913: Globalization, Schooling and Growth', *European Review of Economic History* 1 (2), (1997): 153–190.
7. R. Teijelo, *El Real Conservatorio de Artes, 1824–1887* (Barcelona: unpublished doctoral thesis, Universidad Autónoma de Barcelona, 2011).
8. Riera (1993), Op. cit.; G. Lusa, 'La difícil consolidación de las enseñanzas industriales (1855–1873)', *Documentos de la Escuela de Ingenieros Industriales de Barcelona* 7 (2007): 15–26.
9. J. M. Cano Pavón, 'The Royal Industrial Institute of Madrid (1850–1867)', *Quaderns d'Història De l'Enginyeria* 5 (2002): 66–73.
10. Royal Order of 7/12/1858 about mandatory texts for the industrial engineering schools.
11. Cano Pavón (2001), Op. cit.
12. For the differences between the high theoretical scientific culture in France and the culture of practical science in Britain see Fox and Guagnini (1999), Op. cit.
13. R. Manjarrés, 'Enseñanza de artes y oficios', *La Gaceta Industrial* Nos. 5, 6 and 7 (1888).
14. J. Alcover, 'Las dificultades que ofrece España para el empleo de las máquinas', *La Gaceta Industrial* No. 30 (1865).
15. A. Viguera, *La ingeniería industrial española en el siglo XIX* (Madrid: ETSI, 1961): 15–20; J. M. Ortiz-Villajos, *Tecnología y desarrollo económico en la historia contemporánea* (Madrid: OEPM): 117–8.
16. R. Amengual, *Bielas y Alabes, evolución histórica de las primeras máquinas térmicas a través de las patentes españolas, 1826–1914* (Madrid: OEPM, 2008). See also Notes on Sources (appendix).
17. Fox and Guagnini (1999), Op. cit.

18. P. Sáiz, *Invención, patentes e innovación en la España contemporánea* (Madrid: OEPM): 186–7.

19. US patent N° 866238A.

20. On the notion of elite patenting see Chapter 2 of this book.

21. A. Anduaga: 'The engineer as a "linking agent" in international technology transfer: the case of Basque engineers trained in Liège', *Engineering Studies*, 3 (1), (2011): 45–70.

22. For attempts of periodisation of patenting activity in nineteenth-century Spain, see Ortiz-Villajos (1999), Op. cit, 105–7; and Sáiz (1999), Op. cit., 146–50.

23. D. Pretel, 'The global rise of patent expertise in the late nineteenth century', *Cambridge Working Papers in Economic and Social History* 31 (2017).

24. D. Pretel and P. Sáiz, 'Patent Agents in the European Periphery, Spain (1826–1902)', *History of Technology* 31 (2012).

25. For contemporary accounts of patent agents' activity to commercially exploit inventions see G. Bolibar, 'Misión de los agentes de negocios', *Industria e Invenciones,* No. 7 (14/081908): 61–62; F. Walker, 'Patent Agents and Patent Brokers', *Journal of the Society of Patent Agents* 3 (27), (March 1902): 39–41.

26. For a contemporary account on the high cost and time-consuming nature of the Spanish patent trials see G. Vicuña 'Complemento de la ley de patentes', *La Semana Industrial* year V Vol. V (January 1886): 7 and "Sindicato de inventores", *Industria e Invenciones*, no.139 (28/08/1886): 93.

27. *International Directory of Patent Agents* (London: William Reeves, 1893, 1897 and 1901).

28. Patent applications and assignment contracts for the year 1900 (AHOEPM). See Notes on Sources (appendix).

29. On the concept of professional 'jurisdiction' see A. Abbott: *The System of Professions: An Essay on the Division of Expert Labor* (Chicago: University of Chicago Press, 1988).

30. See, for instance, A. Ungría, 'Representantes y representaciones comerciales', *El Fomento Industrial y Mercantil* No. 355 (10/02/1901): 559–600; No. 356 (20/02/1901): 608–9; and No. 357 (28/02/1901): 615–616.

31. G. Bolibar, 'Publicación de las patentes de invención', *Industria e Invenciones,* No. 156 (25/12/1886): 297–8; G. Bolibar, 'Las asociaciones de agentes de patentes' No. 20 (16/10/1901): 176; G. Bolibar, 'Proyecto de ley sobre propiedad industrial', *Industria e Invenciones*, No. 10 (7/09/1901): 84.

32. G. Bolibar, 'Review of Pedro Estasen, Derecho industrial en España', *Industria e Invenciones*, No. 17 (22/10/1901): 142–3.

33. For the regulation of the patent profession in the United States see K. Swanson, 'The Emergence of the Professional Patent Practitioner', *Technology and Culture* 50 (3), (2009): 519–548; For France see G. Gálvez-Behar, *La République des Inventeurs: Propriété et Organisation de l'Innovation en France, 1791–1922* (Rennes: Presses Universitaires de Rennes, 2008): 171–7; For Britain see H. I. Dutton, *The Patent System and Inventive Activity: During the Industrial Revolution 1750–1852* (Manchester: Manchester University Press, 1984): chapter 5 and A. Guagnini: 'Patent Agents in Britain at the turn of the 20th Century', *History of Technology*, Vol. 31 (2012).

34. Royal Decree of 12/06/1903 (CLE, Nueva Serie, T. xv). Section V on industrial property agents and the register of industrial property agents. Published in BOPI No. 452 (1/07/1902): 1052–4 and in the *Gazeta de Madrid* of 14/06/1903. A translation furnished by Francisco Elzaburu was also published in London by the *Journal of the Society of Patent Agents*, Vol. iv, no. 47–48 (1903): 162–66.

35. Gerónimo Bolibar, 'Observaciones al proyecto de ley de propiedad industrial', *Industria e Invenciones* No.18 (02/11/1901): 158–64.

36. See, for instance, the series of articles by Agustín Ungría celebrating the new regulation and the compulsory requirement to be registered as business agent: Agustín Ungría, 'Los agentes de negocios y la propiedad industrial', *El Fomento Industrial y Mercantil* No. 404 (20/06/1902): 991–2 and No. 407 (20/07/1902): 1016.

37. The remarks of the Royal Order of 22/05/1905 on this issue are eloquent; BOPI, No.453 (July 1905): 837–8.

38. BOPI, No. 461 (November 1905): 72.

39. In March 1906 there were 32 individuals registered (BOPI, No. 470: pp. 423–4); in April 1907, 36 (BOPI 4, No. 96: p. 616) and in June 1908, 37 (BOPI, No. 523, pp. 810–1). See also J. B. Sánchez, *La Propiedad industrial en España: bosquejo histórico y legislación* (Madrid: Instituto Editorial Reus, 1945): 219.

40. Roeb & Co., *Breve resumen de la propiedad industrial* (Madrid: Roeb y Compañía, 1930).

41. Royal Resolution of 17/03/1847. Revised with Royal Order of 25/04/1877. See also P. Madoz: *Diccionario geografico-estadistico-historico de España* (Madrid, 1847): 797–8.

42. Information about business agents' activities can be found in the publications *Boletín de los Agentes de Negocios* (from 1881) and *El Fomento Industrial y Mercantil* (from 1891).
43. For information about the setup of the Asociación de Agentes de Propiedad Industrial see *La Industria Nacional* N° 14 (28/02/1909).
44. Royal Order of 12/05/1909.
45. 'Spain. Industrial Property. Proposed Changes in the Law', *Patent and Trade Mark Review*, V. 12–13 (1913–15): 293–294.
46. RO of 13/12/192, *Gaceta de Madrid*, No. 85 (04/02/1917): 35; No. 88 (07/02/1917): 811; and No. 125 (06/05/1917): 346.
47. 'Rich and Poor Inventors', *Scientific American*, No. 9 (13 May 1854), 277. Cited in R. Thomson, *Structures of Change in the Mechanical Age: Technological Innovation in the United States, 1790–1865* (Baltimore: The Johns Hopkins University Press, 2009): 201.
48. A. Algaba, 'La difusión de la innovación. Las revistas científicas en España 1760–1936', *Scripta Nova* 69 (17), (2000); E. Fernández-Clemente, 'La recepción en España de la Segunda Revolución Industrial: las revistas de ingenieros (1900–1936)', in P. Aubert and J. M. Desvois (eds.), *Les Élites et la Presse en Espagne et en Amérique Latine: des Lumières à la Seconde Guerre Mondiale* (Madrid: Casa de Velázquez, 2001): 171–188.
49. Some pointers on contemporary Spanish monographs focusing on patent issues are: T. Merly, *Legislación industrial española* (1879); F. Lastres, *La propiedad industrial y las marcas de fábrica* (1886); J. Vila, *Manual de patentes de invención* (1902) and P. Estasén, *Derecho industrial de España* (1900).
50. G. Gironi, 'Los privilegios de invención', *La Semana Industrial* (22/01/1886).
51. J. Alcover, *La máquina moderna* (Madrid: Imp. de M. Tello, 1882).
52. For José Alcover and the *Industrial Gazette* see D. Pretel, 'Invención, nacionalismo tecnológico y progreso', *Empiria* 18 (2009): 59–83.
53. A fine literary description of Agustín Ungría's business agency activities in the 1920s in the 'hundreds of official departments' and the origins of this firm in the late nineteenth century can be found in A. Barea, *The Forging of a Rebel* (London: Davis-Poynter, 1972): 390–3.
54. See for example 'Estudios sobre propiedad industrial: Las patentes de invención', *El Fomento Industrial y Mercantil* (20/08/1900).
55. G. Bolibar, *Lo que debe saber el inventor: datos y consejos de utilidad para los inventores o propietarios de patentes* and *Lo que debe saber el que usa*

*marcas: datos y consejos de utilidad práctica.* See also F. A. Lázaro, *Patentes de invención. Instrucciones prácticas* (Madrid, 1895).

56. For a conceptualization and social location of science popularisation in the European periphery see F. Papanelopoulou et al. (eds.), *Popularizing Science and Technology in the European Periphery, 1800–2000* (Aldershot: Ashgate, 2009).

57. Viguera (1961), Op. cit., 61; Amengual (2008), Op. cit., 54.

58. See for example P. Carpentier, *La Loi Espagnole sur la Propriété Industrielle du 16 Mai 1902* (Paris: A. Chevalier-Maresco, 1904); *Ley de propiedad industrial de mayo de 1902* (Barcelona: Administración de Industria e Invenciones, 1902).

59. W. P. Thompson, *Handbook of Patent Law of All Countries* (London: Stevens & Sons, 1882).

60. J. Alcover, 'Convenio internacional para la protección de la propiedad industrial', *La Gaceta Industrial*, No.18 (25/09/1884).

61. G. Bolibar, 'Los derechos del inventor', *Industria e Invenciones* No.102 (2/2/1885): 249–50.

# 4

# The International Dimension

**Abstract** Spain occupied a peripheral position in the globalising Atlantic world economy of the nineteenth century, as evidenced by the way it integrated into the international patent system. Two features characterised the relationship of the Spanish patent system with the broader global dynamics of the time: first, its openness to foreign patenting and strong institutional dependency and, second, the prevalence of transfer agents who facilitated the transmission of knowledge and information to Spain. This analysis of the international dimension of the system concludes with the case study of Julio Vizcarrondo and the Elzaburu agency, the pioneer professional patent business in Spain that assisted foreign inventors and companies.

**Keywords** International patent system • Foreign patenting • Transfer agents • Julio Vizcarrondo

The historical evolution of the Spanish patent system is best understood in its international context. The rise in international connections in the second half of the nineteenth century requires an extension of the scope

© The Author(s) 2018                                                                                     **87**
D. Pretel, *Institutionalising Patents in Nineteenth-Century Spain*, Palgrave Studies in Economic History, https://doi.org/10.1007/978-3-319-96298-6_4

of analysis beyond frontiers and a consideration of the interaction occurring among national systems. The institutional evolution of the Spanish system was reflective of the challenges of the globalising economy. The years 1860–1914 were a period of global expansion of patenting and technology transfers. Inventors and firms from industrial nations extended patent rights to as many national patent systems as possible.[1] This trend was closely related to the expansion of international trade and foreign direct investment. Meanwhile, new industrial technologies were becoming capital-intensive and complex, and therefore harder to transfer, requiring the participation of a variety of transfer agents.

In the historical conjuncture of the late nineteenth century, industrial nations such as the United States, Britain, France and Germany were technological leaders, yielding the highest number of patents, while peripheral-dependent countries, such as Spain, sought to stimulate international patenting in their systems. In other words, the evolution of patent systems in advanced industrial economies and relatively backward countries significantly differed. Yet it was not a simple dichotomy, but rather a mutual interdependence that existed among industrial, industrialising and peripheral nations.

Given that Spain had a smaller market potential and industrial base than other latecomers such as Germany and Italy, patenting in that country was less lucrative to British, French and American capitalists. Although Spain was not a dominant site of patenting for most international inventors, it was one of the preferred national jurisdictions for acquiring patents in the European periphery. As early as 1848, the mechanics' publication *Scientific American*,[2] in its section on inventions and patents, pointed out that 'the foreign laws which afford the highest security for inventors seem to be those of Great Britain, France, Spain, the Roman States and Baviera' and recommended that American inventors extend their property rights to these countries. Some years later, in 1861, the same journal would suggest that American inventors 'bear in mind that, as general rules, any invention which is valuable to the patentee in this country is worth equally as much in England and some foreign country'.[3]

Patent rights registered in Spain flowed overwhelmingly from other countries, particularly Western Europe and the United States.[4] Starting in the mid-nineteenth century, foreigners had sought patents in Spain primarily to secure potential commercial enterprises and advertise their inventions, rather than to manufacture or license their patent rights in this country. International inventors were looking for systems that could offer easy access, high standards of protection and minimal costs. However, the diversity of patent laws and policies posed difficulties for foreign investors willing to acquire patents in several national jurisdictions by reducing incentives for extending protection to countries, such as Spain, at the periphery of industrial development.[5]

## Spain and the International Patent System

Political pressure for policies facilitating international patenting became commonplace from about the mid-1870s. Patent protection limited to national contours was deemed unsatisfactory by those inventors who wished to register their patents in several countries. This international fragmentation elevated transaction costs—including the costs of collecting information—and created legal insecurity. A more convenient legislative uniformity among different patent systems, meanwhile, became a demand of governments, companies and intermediaries. The resolution of heated mid-century disputes on patent protection set the stage for a convergence or at least a greater harmonisation of patent laws during the latter part of the century. The controversy between 1850 and 1875 ended with patent rights advocates as winners.[6] Free-trade supporters—often opposed to national patent regulations— viewed an international patent agreement as tolerable and did not actively resist demands for an inter-governmental arrangement.[7] The ending of the controversies was a turning point that would have long-term consequences, most saliently the emergence of an international patent system.

The international patent system that would emerge in the 1880s had its origins in the growing interactions among national systems, namely European and North American ones. The rising incentives for international patenting had already triggered intricate bilateral treaties of a limited character that dealt mostly with trademarks rather than patents.[8] One such bilateral treaty was the agreement of 1875 between Great Britain and Spain for the protection of trademarks. Similarly, during the 1870s and 1880s, there was a proliferation of transnational agreements, international conferences, supervisory agencies and multilateral institutions on other matters—for instance, the General Postal Union of 1874 and the Berne Convention of 1886 for the Protection of Literary and Artistic Works—all of which Spain signed.[9]

The first step in the creation of a formal international patent system was the Congress on Industrial Property held in Vienna in 1873. This unofficial meeting was attended by inventors, corporations, engineers and lawyers, as well as by chambers of commerce and representatives of several governments. The Spanish government did not send an official delegate, nor did any Spaniard attend in a private capacity. That this congress on patents coincided with the Vienna World's Fair was not happenstance, as new inventions were the main attraction at this kind of international event. The public display of machinery and manufactured products at international exhibitions increased their potential market and attracted financing. Yet these exhibitions simultaneously posed a risk for inventors by opening the doors to an unlicensed use of patented technologies through, for instance, patents of introduction in countries such as Spain.[10] Acknowledging American concerns regarding this issue, the Austro-Hungarian state extended temporary protection to exhibitors at the 1873 world fair. Similar temporary protections had been introduced at the international exhibitions of 1851, 1855, 1862 and 1867.[11] After Vienna Congress, discussions towards an international patent agreement continued with further meetings in 1878 and 1880 in Paris, sponsored by the French government. At the 1878 International Congress, the Spanish delegation was headed by the politician José Emilio de Santos, Spanish commissioner of the world's fair of 1878. For the 1880 conference, Spain did not send any delegates.

Finally, in 1883, at the International Convention for the Protection of Industrial Property in Paris, a formalised international patent system was established.[12] The result was an international agreement for a gradual convergence of national patent laws among 11 nations, including Belgium, France, Italy and Portugal.[13] Spain was also a founding member. Britain joined in 1884, the United States in 1887, Japan in 1899 and Germany in 1903 (see Table 4.1 for a summary of the years of accession to the 1883 Paris Union for several major countries).[14] Spain was represented at the Paris Conference by the industrial engineer Felix Márquez López, who had been a professor of mechanics at the Royal Industrial Institute of Madrid and at the time was Director of the Royal Conservatory of Arts, the institution in charge of granting patents in Spain. Márquez

**Table 4.1** National patent laws and year of accession to the Paris Union

| Country | First patent law | Accession to 1883 Paris Convention |
|---|---|---|
| Great Britain | 1624 | 1884 |
| United States | 1790 | 1887 |
| France | 1791 | 1884 |
| Austria | 1810 | 1909 |
| Germany | 1815 | 1903 |
| Belgium | 1817 | 1884 |
| Spain | 1826 | 1884 |
| Mexico | 1832 | 1903 |
| Sweden | 1834 | 1885 |
| Portugal | 1837 | 1884 |
| Italy | 1859 | 1884 |
| Canada | 1869 | 1923 |
| Turkey | 1879 | 1925 |
| Japan | 1885 | 1899 |
| Switzerland | 1888 | 1884 |
| Australia | 1903 | 1907 |

was a technical delegate, not a politician, and therefore could not ratify Spanish adhesion before consultation with the Spanish government.[15]

The agreement of 1883 is considered one of the first multilateral treaties in history. This treaty enshrined the doctrine of the natural law of invention, the principle of national treatment and the right of priority. Its main achievement was the establishment in 1884 of the Union for the Protection of Industrial Property, headquartered in Bern. The Paris Union was created with the aim of coordinating a gradual harmonisation of patent laws among signatory countries. This agreement eased international patenting and reduced complexity, facilitating an international market for technology and rights. Issues such as international priority rights, the concept of novelty required for a valid patent, compulsory working, registration and publicity were settled.[16] The Paris convention and subsequent meetings did not introduce a simple and centralised register of patents for all countries, nor did they reduce costs for multiple international patenting. The convention contained few standards of substantive law and implicitly accepted the inevitability of national diversity by embracing the principle of non-discrimination on the basis of the nationality of the applicant.

The instability and imperfection of the international system allowed for institutional diversity, particularly policies that facilitated technological emulation at a low cost.[17] Although the Paris Union facilitated the coordination of the signatory countries, it did not eliminate the sovereignty of national patent policies. The Paris Union did not mean in practice the elimination of Spanish autonomy on the matter, as bureaucratic practice, legal standards and the interpretation of laws left room for divergent organisational arrangements. What it encouraged, if anything, was interconnections with other national patent systems. This, in turn, meant growth in the dimension and complexity of patenting activity in Spain. The interconnections also deepened the engrained institutional features and earlier tendencies of the Spanish system, such as its openness, dependency and peripheral position.

It would be inaccurate to adopt a non-statist approach to the understanding of the making of the international patent system. The Paris Union was primarily a political project of nation-states. National systems still provided the necessary infrastructure for the international system. The Paris agreements had the effect of consolidating strong national patent models, with domestic patent regulations and policies remaining crucial. National patent institutions grew more interdependent while continuing to maintain their heterogeneous character. On a practical level, it was an asymmetrical system, with marked differences in the levels of patenting activity in each country. The persistent diversity within patent cultures at the turn of the twentieth century resulted from different levels of industrialisation. The aspiration towards a 'global patent' of international scope obtained through a single procedure of registration was far from achieved. The international patent system remained fragmented, and demands for more uniformity among national laws would continue.[18]

The provisions agreed upon in 1883 slowly entered into Spanish legal system and had some practical consequences on the administration and publicity of patents. Although Spain was a founding member of the Union, the transposition of the agreement to Spanish law was slow. This would cause, during the following years, complaints from other countries, the Spanish press and even Spanish civil servants working on industrial questions. It was not until 1886 that, in response to the Paris Union, a specialised official publication, the Bulletin of Intellectual and Industrial Property, was set up. Following international agreements, Spain also regulated industrial models and designs.[19]

The Paris Convention was followed by conferences of revision. The first one, held in Rome in 1886, served to consolidate the Paris Convention but had very limited substantive amendments.[20] Subsequent conferences of revision took place in Madrid (1890) and Brussels (1897 and 1900). The Madrid Revision Conference for the Protection of Industrial Property took place in April 1890 at the San Fernando Real Academy of Fine Arts. The official language was French, and each country had a single vote. When the conference took place, Spain was in a period of relative political stability and industrial take-off. For the Spanish government, the organisation of this major diplomatic gathering was an opportunity to show the international community not just its commitment to the Paris Union but also its respect for patent rights. Segismundo Moret, who was president of this conference, declared at the first plenary session that Spain had reached a degree of industrial progress 'to guarantee not only the rights of Spaniards but also those of foreigners in their economic and industrial relations with Spain'.[21]

Thirteen countries sent delegates to the Madrid meeting, including France, the United Kingdom, the United States and Italy. The head of the German patent office also attended, although Germany was not yet a member of the Union. Spain, besides being the organiser, had four delegates representing Spanish interests: Segismundo Moret (a member of the Spanish Parliament and former minister of Foreign Affairs), the Count of San Bernardo (Director of Agriculture, Industry and Commerce and also a member of the parliament), the industrial engineer Enrique Calleja (Head of the Spanish Patent Office) and Luis Mariano de Larra (Director of the Spanish Bulletin of Intellectual and Industrial Property).[22] Larra had also been the Spanish representative at the Conference of Revision in Rome in 1886. Another Spaniard, Luis Prota, an official of the Ministry of Public Works (Ministerio de Fomento), was one of the secretaries for the Madrid conference. The United Kingdom was represented, among others, by the comptroller general of the British Patent Office, Henry Reader-Lack.[23] Delegates at the Madrid conference discussed several protocols, particularly concerning the interpretation of the Paris Convention, the administration of the International Bureau, the international registration of trademarks and the repression of false indications of origin. The final agreements concerning patents were, however, limited in scope.

The appointment of Spanish delegates at the Madrid Revision Conference precipitated some criticism among the Spanish newspapers, which saw the Spanish representatives as politicians with limited experience and practical knowledge about industrial matters. The benefits of the Madrid Conference for Spain were also questioned. For instance, an editorial in the newspaper *El País* argued that this meeting was not going to have any practical consequences for Spanish industry, as legislation on patent rights was inefficient in encouraging national factories and industries.[24] Other newspapers, such as *El Imparcial*, were enthusiastic about the Madrid Conference and the Spanish commitment to intellectual property protection and saw such an event as an important opportunity for Spain's economic modernisation.[25] In 1891, a year after the Conference of Revision, another diplomatic meeting was held in Madrid, where a protocol concerning trademarks was ratified.

# Foreign Patenting and Technology Transfer

During the second half of the nineteenth century, there was a continuous flow—albeit not without certain obstacles—of technology, capital and qualified personnel from Western Europe and the United States into Spain's textile, metalworking, mining, chemistry, railways, merchant navy and paper industries.[26] Technology transfer mechanisms to Spain varied considerably throughout the nineteenth century, from formal to relatively informal. The major technology transfer mechanisms were trade, the migration of industrial experts, foreign direct investment and the reception of technical literature. Patenting served primarily as a mechanism of communication of property rights, although socio-technical networks built around the Spanish patent office were a complementary path of transmission of technological knowledge.[27]

Between 1878 and 1907, foreign patenting (including patents of importation) reached a level of approximately 70% of total patenting in Spain (see Fig. 4.1). Spanish industrial take-off during the first phase of the Restoration regime (1875–1902) was built on foreign technologies emerging from the industrial leaders of the time (France, Germany, United Kingdom and the United States). This occurred in most industries, most saliently the advanced sectors associated with the Second Industrial

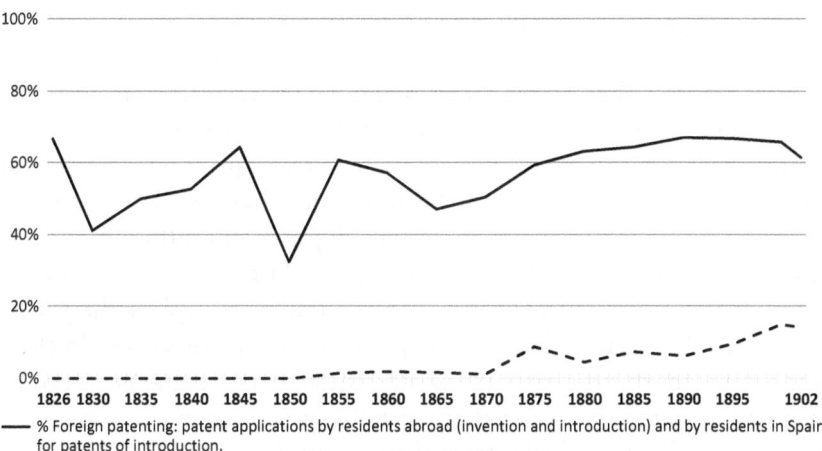

**Fig. 4.1** Foreign patenting in Spain, 1826–1902

Revolution, which included electricity, chemicals, gas, mechanical equipment, transports and iron.[28] The increase in foreign patenting in the period 1878–1902 did not necessarily mean higher levels of technological adoption or dissemination nor the development of domestic technical capabilities. It seems clear that the direct transfer of technology to Spain through the patent system was relatively limited during the late nineteenth century. Several technologies were patented in Spain but not always diffused and generalised at the same speed as in other parts of Western Europe. As in other peripheral markets, foreign patenting was primarily a defensive strategy to complement the commercialisation of technology.[29]

The standardisation of technical specifications in industrial sectors, such as plastics, electricity and heavy machinery, facilitated the registration and exchange of property rights of invention among different countries. This may explain why during the second half of the century and the first decades of the twentieth century some of the most economically valuable foreign complex technologies, from production tools to capital-intensive railway installations, were registered in the Spanish patent system. Professional inventors associated with the Second Industrial Revolution, such as Henry Bessemer, Thomas Alva Edison, Guglielmo Marconi, Alfred Nobel and William Thomson, patented in Spain.

Strategic partnerships built around patent agreements were a common mechanism of technology transfer during the second half of the nineteenth century. Transnational collaborations between foreign inventors and Spanish industrialists, often mediated by a variety of intermediary agents, reduced the transaction costs of foreign patent lodgement in Spain. This kind of mechanism for technology transfer also reveals the severity of long-term constraints to the adoption and diffusion of technology in Spain and explains the need for industrialist to resort to the assistance of professional expertise. An illustrative example is the patent agreements between the Basque steel-producing company Ybarra Brothers and foreign inventors during the 1850s. The patents that the Ybarra company was granted on steel-making technologies are worth investigating. In this case, the individual intermediating with foreign inventors was the manager of the Ybarra company, José de Vilallonga. In 1854, for example, Vilallonga, Ybarra and the French mining engineer Adrien Chenot entered into a collaboration for the exploitation of Chenot steelmaking inventions in Spain.[30] Later on, in 1856, Ybarra Bros., Vilallonga and Chenot were jointly granted a patent for an indirect iron-making process consisting of sponge furnaces.[31] Several of these iron furnaces were installed, with the assistance of Chenot, in Ybarra's metallurgy plants.

In 1856, Vilallonga also agreed to a £5000 patent licence with the English industrialist and engineer Henry Bessemer for the exclusive use of his process in Spain for the mass production of steel from wrought iron without fuel.[32] A few months later, in January 1857, the firm Bessemer & Longsdon—established by Henry Bessemer—obtained a Spanish patent of invention for 15 years and assigned it to the firm Ybarra Brothers.[33] Bessemer had obtained a British patent for the same procedure in 1855, but he was covered only within British territory, not in Spain.[34] Bessemer's invention was already well known by advanced ironmasters throughout Europe; on 11 August 1856, he had presented a paper explaining his invention at the meeting of the British Association for the Advancement of Science, published in *The Times* a few days later. Bessemer had devised a new concept for producing stronger iron by blowing air into an ovoid open-topped furnace—the converter—using molten pig iron. The mild steel obtained using the Bessemer method was not only stronger but also far cheaper and thus ideal for mass industrial production. Soon Bessemer began to grant licences for his procedure to various iron companies in Britain.

Just a few weeks after Bessemer's public presentation of this invention, Ybarra Bros. and José de Vilallonga were granted a five-year Spanish privilege of introduction for the Bessemer process.[35] They were not the inventors of this new procedure, but Spanish patent law protected the emulation of foreign technological improvements. The documentation Vilallonga presented to the Conservatory of Arts included a concise technical memorandum on the new steel converter, based on explanations from Bessemer's article. In the patent application document, Vilallonga was confident that with the use of the Bessemer system, Spanish steel production would 'reach unimaginable levels'. Although Ybarra and Vilallonga had already obtained the Spanish patent rights for the Bessemer invention in 1855, they decided to sign a patent licence agreement with Bessemer in 1856 to receive technical assistance and—given that patents of introduction could only last five years—extend the duration of the monopoly to 15 years. The Bessemer invention was successfully implemented in 1858 by Vilallonga and the Ybarra company in Santa María de Guerizo, historically a major centre of iron mining and manufacturing in Spain. Bessemer himself was present at this first trial and even supervised the installation of the converter. Although the trial of the Bessemer system in Spain was a success, this steel converter still presented major technical challenges,[36] and most of the first trials in other parts of Europe failed. Vilallonga also received a confidential report from Bird & C., his agent in London, warning about the converter's technical snags. The major problem was excessive oxidation, which resulted in fragile iron ingots. Among the first trials in continental Europe, only the attempts of Vilallonga and Bessemer's Swedish patentee Göran Göransson (in July 1858) were successful, surely due to the propitious chemical composition of the iron mineral used. However, while the introduction of this invention to Spain remains a mere historical curiosity, Göransson identified and corrected the problems of Bessemer's invention, thus succeeding in producing and commercialising steel.

Vilallonga and the Ybarra firm finally withdrew from the Bessemer enterprise, despite the considerable investment they had made. They decided to continue using Chenot sponge furnaces, a French process for producing steel that had less long-term potential but was more reliable. Ultimately, Vilallonga's decision to prolong the use of the indirect iron-making methods was, unquestionably, a mistake and a missed opportunity. Beginning in the mid-1860s, the Bessemer method yielded substantial productivity gains and was rapidly diffused.[37] It became, for the last third of the century, the dominant process for steelmaking in Europe and its importance for the shipbuilding and railway construction industries was remarkable. The Bessemer converter was not reintroduced to produce steel in Spain until 1885. It was in that year that the Basque firm Altos Hornos, again owned by José de Vilallonga, became the first Spanish company to produce and commercialise steel in Spain using the Bessemer process.[38]

The episode of the Bessemer converter's early arrival in Spain speaks to the drawbacks of placing too much emphasis on entrepreneurs' attitudes when studying the dynamics of international technology transfer. Henry Bessemer blamed Vilallonga for not having enough entrepreneurial spirit.[39] Bessemer also had complaints about British businessmen's conservatism towards technology. For instance, Bessemer himself relates in his autobiography that at the meeting at Cheltenham where he presented his paper, the British iron masters saw his invention 'rather as a good joke than a reality'. However, the truth is that Vilallonga, like many industrialists elsewhere, was forced to choose between available technologies. The demand for iron production in the Spanish market was too low to allow for plants that used different methods. Given the restrictions of the Spanish market, the technological choice depended on the availability of experts with experience in the iron and steel industries, as well as on the quality of the production and the reliability of the method. In other words, the decision to abandon the Bessemer converter was motivated not only by the relatively high economic cost of the investment but also for technical reasons. A large part of the problem lay in the empirical nature of Bessemer's inventive process. Bessemer was a practical man lacking scientific training, and he did not fully understand the chemistry of the process he had invented. This seems clear when we look at Bessemer's obituary in *The Engineer*, which stated that 'it is nearly certain that Bessemer never mastered the chemistry of his steelmaking process'.[40]

Furthermore, the internationalisation of the Spanish patent system was closely related to the increasing corporate control of valuable patents in late nineteenth-century Spain. Although many companies remained reluctant to extend their rights to Spain, several large companies extended patents to Spain during the 1880s and 1890s. In 1902, patenting from foreign companies reached over 14% of total patenting in Spain, whereas in 1860 this figure had been only 2% (see Fig. 4.1).[41] Large foreign companies including Krupp, American Bell, Vickers Ltd., Siemens-Halske, General Electric and Schneider Cie. patented in Spain during the late nineteenth century. Foreign patenting was particularly apparent in firms created around the consolidation of patent rights such as Babcock Wilcox and the various Edison companies. The economic value of the inventions protected by foreign companies in Spain was higher than those protected by independent inventors. Many of these corporate patents were in advanced sectors such as electrical engineering, chemicals, armament, communication and machine tool industries, whose research and development costs were higher. Patents obtained by foreign corporations had a longer duration and higher rates of implementation and assignments than average patents registered in Spain. The number of patent assignments and licences by foreign companies increased from the 1880s, although it remained lower than in France and Britain and substantially lower than in the United States. Contracts of assignment and licences of patents of foreign companies were overseen by Spanish businessmen, lawyers or agents.

There were strategic reasons for the assiduous lodging of patents in Spain by large foreign firms during the late nineteenth century, although the strategies were industry-specific. In general, the primary strategy of manufacturing companies was to reduce potential competition. Protective patenting in many national jurisdictions served to limit imitation. The threat of litigation was used by foreign companies to preserve their market position at the international level. Firms also used this defensive strategy in other countries of the economic periphery as well as in oligopolistic markets of advanced industrial nations.[42]

Foreign industrial firms that had obtained patents in Spain remained reluctant to use them in the country itself at an industrial level. In Spain the cost of foreign direct manufacturing linked to patenting technologies was high, although there were differences depending on the sector and the country of origin of the firm.[43] Even when these firms did not directly exploit their inventions, their patenting activity had enhanced the growth of complementary businesses. Among them were a number of engineering consulting firms working on patent issues, specifically in electrical technology and chemical dyestuffs. Foreign companies also had corresponding subagents in Spain to communicate their industrial property rights.

Only a few firms, including the Edison and Bell companies, systematically licenced the rights of their Spanish patents. Similarly, only a minimal group of companies established branch plants and subsidiaries to exploit their property rights in Spain during the last two decades of the century. An example of the latter was the British engineering firm Vickers, which had a subsidiary in Spain (The Placencia de las Armas Company) from 1897 to produce armaments, thereby avoiding protectionist policies and exploiting the patent rights it had obtained in Spain.[44] Most foreign corporations, such as Babcock & Wilcox, Krupp and the Westinghouse Electric Company, engaged in the direct trade of technology once they had secured their patents in Spain. These companies often provided their Spanish clients with the necessary services for setting up and operating the new technology, including additional equipment, training and expert personnel.

The internal correspondence on patent management among the various Edison companies between 1889 and 1898 reveals that persistent patenting activity in the major European countries—and indeed peripheral ones as well—was a priority strategy of large American firms.[45] The Edison Electric Light Company of Europe, for instance, whose total capital of $2 million had been garnered entirely from patents,[46] was created in 1880 with the express purpose of taking out patents related to electric lighting in European countries, excluding Britain. As the firm's documentation reveals, the only activity of this company was the management of Edison's patents: '[the company is] not engaged in manufacturing, nor does it on lease, or use any factory, machinery or manufacturing appliances. It is not engaged in selling machines or merchandise of any description, nor does it carry a stock of goods or merchandise.' The patenting activity of this company was limited to Austria, Hungary, Russia, Italy, France, Belgium, Denmark and Spain (the Spanish colonies excepted).[47] In their writings, Edison's biographers, Frank Dyer and Thomas Martin, stated that Edison and his companies were granted 54 patents in Spain. This contrasts with an 1892 report by Edison's legal adviser, Sherburne B. Eaton, that Edison had 61 patents in Spain and its colonies. In reality, according to patent records, as many as 84 patents were granted in Spain to Thomas Alva Edison and several of his companies between 1878 and 1903.[48]

# Transfer Agents

Nowhere was the impact of international dynamics more apparent that in the ubiquity of transfer agents in the Spanish system. From the 1870s, different actors actively engaged in the transmission of patent rights to Spain, while simultaneously acting as knowledge brokers and channels for the circulation of information related to the world of patents. Spanish engineers, lawyers, industrialists and commission agents became intermediaries of foreign agencies, inventors and firms based in New York, Washington, DC, London and Paris, just to name a few major patenting centres.[49] This situation was not only evident on patent registers but was also pointed out by the specialised press of the day. The industrial engineer Teodoro Merly, for instance, in various articles published in the daily *El Liberal* in 1890, called attention to the rise of new expert professionals intermediating in technology flows and patenting to Spain from abroad.[50]

During the 1850s and 1860s, general purpose commissioning houses had dominated the field of international patent intermediation in Spain. Among the most active in the provision of international patent services were the lawyers Juan and Leopoldo Barrié, based in Madrid. The Barriés were correspondent subagents of professional intellectual property agencies based in other European cities, particularly London and Paris, who obtained numerous patents between 1853 and 1878. They worked mainly as correspondents for the French patent agency and consultant engineering firm *Armengaud Aîné*, established as early as 1836, which was connected with the inventors' magazine *Le Génie industriel*.

The leader of the field from the 1850s to the mid-1870s was the merchant house C. A. Saavedra (also called Saavedra & Riberolles), often subcontracted by Brandon and Morgan-Brown (engineers and patent agents based in London) or Charles Thirion (a French civil engineer and patent agent). The Saavedra & Riberolles commission house had offices in Madrid, London and Paris. Apart from channelling hundreds of foreign patents to Spain, this firm provided international trade services and commercial representation to foreign merchants, manufacturers, authors and publications. For instance, this house advertised and distributed more than 90 foreign newspapers and journals like *The Times* and *Le Messager de Paris*. It even assumed the function of a travel agency and edited the tourist guide *Manual del Viajero Español* (Spanish Traveller's

Manual). Two business agents working for the Saavedra firm—Domingo and Telesforo Algarra—specialised in patenting intermediation.

It was from the late 1870s that the overwhelming majority of foreign inventors patenting in Spain began resorting to professional intermediaries. International patenting to Spain expanded during that decade, becoming a business in its own right. Three Spanish agencies lodged most of the foreign inventions introduced into the Spanish system during the 1880s and 1890s. The first two—Vizcarrondo (later renamed Elzaburu) and Clarke & Modet—were based in Madrid and run by lawyers. They concentrated on administrative duties, providing legal assistance though usually not direct support for the technical aspects of the technologies. These two agencies were for the most part subagents or correspondents of agencies based in industrial countries. A third agency, Oficina Internacional de Propiedad Industrial, was an engineering consultancy established in 1884 in Barcelona by the industrial engineer Geronimo Bolibar and connected with the magazine *Industria e Invenciones* (Industry and Inventions). These were not the only players, but they were the most important. Some agencies also assisted Spanish inventors in obtaining protection for their inventions in foreign systems, although this service only represented a minor part of their activities.

International agents translated and adapted foreign patents to fulfil the requirements of Spanish regulations.[51] Also, the non-textual parts of patents, like the models and technical drawings, were often substantially modified when transferred. For example, in 1898, Edison's patent lawyers sent detailed instructions on the securing of patent rights for an ore-milling apparatus in Spain, among other countries.[52] The lawyers Francisco Elzaburu and Gonzalo Pelligero, the correspondent agents in Spain, were requested to obtain six Spanish patents for this Edison's invention.[53] In the international correspondence and the letters exchanged with Elzaburu and Pelliguero,[54] Edison's expert in-house agents were especially concerned about the quality of the Spanish translation of the details of Edison's invention for concentrating magnetic iron ores. While they considered Pelligero a 'very careful and trustworthy man' and were satisfied with his work as an intermediary agent, they nonetheless did ask him to supervise a second more accurate translation. Edison's lawyers also expressed their willingness to have a final look at these Spanish patent applications before they were submitted. In the case of this company during the 1880s and 1890s, Edison's in-house lawyers coordinated international patenting, reducing transaction costs within this multinational firm—particularly information costs. However, external intermediaries were additionally hired in each national jurisdiction. The difficulties involved in coordinating patent control—and eventually transnational research, production and commercialisation—within multinational companies largely explain the rise in international intermediaries and corresponding subagents.[55]

An important part of the activity of agents working for foreign patentees was drafting licences and providing counsel on property rights assignments. The Italian inventor and industrialist Guglielmo Marconi was one of the patentees assisted in these matters by intermediaries in Spain. For instance, the lawyers Alberto Clarke and José Gómez Acebo of the Madrid-based agency Clarke, Modet & Co assisted Marconi in 1896 and 1897 in obtaining, transferring and verifying the working of his patent on wireless technology. Following instructions from the Carpmael & Co's London-based patent firm, Clarke, Modet & Co. in 1896 secured Marconi's Spanish patent on this technology, which would go on to become a worldwide commercial success. A year later the same agency assigned Marconi's Spanish patent to London's Wireless Telegraph and Signal Company Limited of London and certified that the actual implementation of the invention was performed following the requirements of the Spanish law.[56] Marconi also obtained patents for his famous invention in Britain, France, Germany, Spain, Italy, Russia, the United States and India.

In the reformed system post-1878, international networks of intermediaries reduced the transaction costs incurred by foreign patentees, such as information and travelling expenses. Expert networks increased security, lowering uncertainty among foreign patentees. The effective lodging of patent rights in Spain was upheld by intermediation, although foreign patenting continued to be reserved for those with enough economic resources to afford the costs of these professionals. Agents charged high fees, and the chain of intermediaries required for patenting in Spain increased the standard expenses of registering (i.e. official fees) considerably. For example, according to British agents Henry Johnson and James Johnson, in 1866 the intermediation charges for obtaining a patent in Spain ranged from $20 to $30, in addition to a government fee that ranged from $2 to $12 depending on the duration and type of the patent.[57] It seems clear that agent fees continued to be high during the late nineteenth century. Even when information costs were reduced by transfer agents during the last decades of the nineteenth century, the majority of foreign inventors were reluctant to routinely extend their patents to Spain.

From the 1880s several Spanish agents were active members of foreign professional associations such as the *Chartered Institute of Patent Agents* and the *Syndicat des Ingenieurs-Conseils en Matiére de Proprieté Industrielle*. In 1902, five Spanish agents and lawyers, such as José Pellá and Barrasa and Breuer, were also members of the *International Association for the Protection of Industrial Property*, an interest group established in 1897 with the aim of promoting an effective implementation of the provisions of Paris Convention.[58] These professional associations had foreign as well as local members, thus facilitating patenting activity in the industrial periphery. Associations were also critical to patent agents' international lobbying efforts and an instrument that promoted international cooperation. Through these associations, Spanish agents established collaborations with foreign patent experts, garnered reputation and exchanged valuable information about professional practices.

Several Spanish agents were nominated by the *Chartered Institute of Patent Agents* (CIPA) council and elected as members during the last two decades of the nineteenth century. The Spanish agents Julio Vizcarrondo (from 1882), Alberto Clarke (from 1884), Gerónimo Bolibar (from 1887), Fernando Modet (from 1892) and Francisco Elzaburu (from 1894) were all active foreign members of this association. Bolibar and Elzaburu were also active in the *Institute of Patent Agents*, a cosmopolitan institution established in London in 1893 with dozens of foreign members. Following the recommendation of CIPA, the two leading Spanish agents in international patenting, Francisco Elzaburu and Alberto Clarke, repeatedly urged the Spanish government to revise Spain's industrial property law. Abuses related to the granting of patents of introduction, delays in patent assignments and the strictness of working clauses elicited foreign agents' ardent criticism.[59] For instance, in June 1888 members of the CIPA urged Julio Vizcarrondo and Alberto Clarke to become 'armed with the views of the Institute' in their negotiations with the government and suggested that the two agents call for changes in Spanish law that would grant stronger rights to foreign patentees.[60] A couple of years earlier, in 1886, Vizcarrondo had sent a letter to the same institute stating that, as a recently elected member of the Spanish parliament, he would take every opportunity to amend the provision and interpretation of Spanish patent law, especially as they pertained to the working clauses of the patents and the duration of patent protection.[61]

## Julio Vizcarrondo and the Elzaburu Agency

Spanish agents left few comprehensive historical records of their activities during the nineteenth century. The original business diaries and registration books of patents and trademarks of the Puerto Rican lawyer Julio Vizcarrondo from 1875 to 1888 are an exception. In 1865 Vizcarrondo founded the Anglo-Spanish General Agency and Commission House, the first Spanish professional agency to work for foreign firms and inventors. Although this account is limited insofar as this agency worked almost exclusively in international patenting and trademark activities, the case does shed light on the operations of Spanish agents in international socio-technical networks. The Elzaburu firm—which remains one of the largest agencies in international patenting and trademark application in Spain, Portugal and Latin America—had it is origins in Vizcarrondo's business. Between 1875 and the late 1880s, Vizcarrondo was the leading Spanish patent expert, acting as the agent for approximately 20% of the total patent applications in the Spanish system, including inventions by Thomas Alva Edison, the Bell Telephone Company and the German steel producer Krupp.

Vizcarrondo (1829–1889) was a prominent and multifaceted personality. Lawyer, liberal politician, philanthropist and journalist as well as a leader of both the Spanish slavery abolitionist movement and the Spanish evangelical community, Vizcarrondo was educated in San Juan, Paris and Madrid and obtained his law degree from the Universidad Central de Madrid. Because of his political stand against slavery in the Caribbean, Vizcarrondo was exiled in 1850 to New York City, where he met and married Harriet Brewster, daughter of Henry Brewster Stanton, a leader in the American abolitionist movement. Vizcarrondo's international education and four years of exile in New York would give him the ideal personal and professional background to build, late in his life, a sizeable patent agency in Madrid. In 1854 he returned to Puerto Rico (then a

Spanish colony), where, in addition to his abolitionist and charitable activities, he worked as a lawyer and helped to promote industrial activities on the island. Also in Puerto Rico, he edited the publication *El Mercurio,* which he founded in 1857, and wrote several books, including a history book in 1862, *Elementos de Historia y Geografía de Puerto Rico* (Elements of History and Geography of Puerto Rico) and a maths book in 1863, *Tratado de Aritmética* (Treaty of Arithmetic). Before moving to Madrid in 1863, Vizcarrondo worked mainly as a representative for American and British merchants in Puerto Rico and Cuba.[62] In Madrid, he continued his journalistic endeavours, including the establishment of the periodicals *Revista Hispano-Americana* (Hispanic-American Magazine) and *El Abolicionista Español* (The Spanish Abolitionist).

Although Vizcarrondo had founded his business agency in Madrid in 1865, the firm did not come to specialise in the patent realm until a decade later. From 1875, Vizcarrondo began providing a wide range of professional patent and trademark-related services for international inventors and firms.[63] Soon he would become a professional intermediary with expert knowledge in intellectual property rights. With the help of his wife, Harriet Brewster, who was fluent in several languages, Vizcarrondo managed to build an industrial property business that enjoyed intensive international activity from the start. Vizcarrondo was not only the foremost Spanish patent agent of the 1870s and 1880s but also a leading agent in trademarks, just after the lawyer Rafael Hacar of Madrid.

Vizcarrondo was also the first Spanish agent to become a member of foreign professional associations such as the French *Syndicat des Ingenieurs-Conseils en Matiére de Proprieté Industrielle* and the British *Chartered Institute of Patent Agents.* In the late 1880s, Francisco Elzaburu (1862–1921), Vizcarrondo's nephew and himself a civil lawyer educated at Madrid's Universidad Central, joined the firm. After Vizcarrondo's death in 1889, Elzaburu took over his uncle's position as director of the industrial property business. Elzaburu would soon become the most influential Spanish patent agent, one who was extremely active in national and international associations.[64] At the turn of the century, Elzaburu embarked on a series of lobbying endeavours with the goal of institutionalising patent intermediation in Spain, following earlier experiences in other countries. With Elzaburu's leadership, this agency would grow in dimension, hiring other experts and absorbing other agencies, such as the business of Eladio Pomata in 1905.[65] From 1897 until 1913 Francisco Elzaburu was also a member of the executive committee of the International Association for the Protection of Industrial Property and from 1907 president of La Asociación Española de Agentes de la Propiedad Industrial y Comercial (The Association of Industrial and Business Agents).

A study of Julio Vizcarrondo's business diaries from 1875 to 1888 yields some tentative conclusions about the functioning of patent intermediation in Spain, and more broadly about the instiutionalisation and organisation of the Spanish system during the late nineteenth century. From its creation by Vizcarrondo, this agency worked almost exclusively as a Spanish representative of foreign firms and inventors; Vizcarrondo himself always worked as a subagent or correspondent for American, French and British patent agents. In this period, Vizcarrondo worked for more than 20 agencies, the most relevant being Henry Johnson (London), Haseltine, Lake & Company (London), Charles Thirion (Paris), Knight Brothers (Washington), Bletry Frèrès (Paris), Charles Desnos (Paris) and Munn & Co. (New York). Foreign lodgements in Spain channelled by Vizcarrondo's patent firm from 1875 to 1888 originated in London, Paris, New York, Berlin, Vienna and Washington, DC, the era's dominant patenting cities and administrative centres, where international patent agencies and professional associations of agents were located in close proximity to patent offices. Vizcarrondo's foreign counterparts were all large agencies whose associates were active members of the various associations of patent agents in which Vizcarrondo participated. Two agents, Charles Thirion and Robert W. Lake, were Vizcarrondo's principal foreign counterparts during this period. Lake, a draughtsman, flooded hundreds of American and British inventors' patent portfolios through his London-based international firm Haseltine, Lake & Co, including, for example, a 20-year patent of invention in 1886 to the Shipman Engine Export Company of Boston for an improvement in steam engines.[66] In the trademark business, Vizcarrondo's most relevant counterpart was the agent Joseph Seymour Salaman, a renowned lawyer and solicitor based in London who specialised in trademark applications and disputes. Salaman was a well-known expert and author of the specialised publication *Trade Marks: Their Registration and Protection in the United Kingdom and Abroad* (1876).[67]

An important part of Vizcarrondo's work was the extension of patent rights to Portugal, for which he subcontracted the Lisbon-based agent C. Pedroso, thereby substantially increasing the costs of the operation. Vizcarrondo extended patents of Thomas Alva Edison and the International Bell Telephone Company to Portugal. By contrast, Vizcarrondo did not, however, extend property rights to other Latin American countries, aside from the Spanish Colonies in the Caribbean, because, as he acknowledged in a July 1885 letter to Haseltine and Lake, there was not an agent there that he could trust. For the Spanish Caribbean colonies, he intermediated for European machinery manufacturers. For example, during the 1880s Vizcarrondo represented the Scottish firm Duncan Stewart and Co. in its application for Spanish patents. In reality, Vizcarrondo acted in these operations as a correspondent of the British agent Henry Johnson. Following Johnson's instructions, Vizcarrondo arranged the translation of technical specifications and the adaptation of the mechanical drawings. Vizcarrondo also certified the working of the patents, following the provisions of Spanish law.

International agency activity was very lucrative for Vizcarrondo, but only to the extent that he had a trustworthy counterpart abroad. Although the scales of Vizcarrondo's tariffs for procuring patents and for other services varied considerably depending on the number of intermediary agents flooding the patent portfolios, his rates were for the most part extremely high. His fee typically amounted to four or five times the official fee charged by the Spanish patent office. For his long-term foreign correspondents such as Haseltine, Lake & Co., his fees were slightly less expensive, and trusted foreign agents were charged not for individual operations but for groups of operations or on a monthly basis. Vizcarrondo did not have a fixed charge for each service but fees were negotiated periodically.[68] The amount of time and effort required to prepare the translations and working clauses determined the value of patent and trademark services, which was more significant in some cases than in others. Vizcarrondo did not charge foreign inventors directly but instead acted as a subagent of foreign patent agents. It seems that Vizcarrondo had permanent agreements with his foreign counterparts concerning the costs of his services. The bulk of his fees stemmed from translation services, for which he charged by the page. He charged additional fees to advertise the patent and to provide state-certified proof of patent implementation, and his invoices included a certain percentage of patent and trademark sales as well as a royalty for each assignment. Rights extensions to other countries and the overseas territories drew additional fees.

# Notes

1. I. Inkster, 'Patents as Indicators of Technological Change and Innovation – An Historical Analysis of the Patent Data 1830–1914', *Proceedings of the Newcomen Society* 73 (2003): 179–208.
2. 'Patent Laws', *Scientific American* 15 (3), (01/01/1848): 118.
3. The magazine *Scientific American* often reminded American inventors that the agency Munn & Co., editor of this publication, 'obtain[s] patents in Great Britain, France, Belgium, Prussia, Austria and Spain, with promptness and dispatch'. 'Instructions about European Patents', *Scientific American* 2 (1005), (13/07/1861): 3.
4. J. M. Ortiz-Villajos, 'Spanish Patenting and Technological Dependency, pre-1936', *History of Technology* 24 (2002): 203–32; P. Sáiz, 'Los orígenes de la dependencia tecnológica española. Evidencias en el sistema de patentes (1759–1900)', *Economía Industrial* 343 (2002): 83–95; D. Pretel, 'La economía política del sistema español de patentes en perspectiva internacional, 1826–1902', *Investigaciones de Historia Económica* 13 (3), (2017): 190–200.
5. C. May and S. K. Shell, *Intellectual Property Rights: A Critical History* (London: Lynee Rienner Publishers, 2006): 111.
6. F. Machlup and E. Penrose, 'The Patent Controversy in the Nineteenth Century', *The Journal of Economic History* 10 (1): 1–29.
7. May and Shell (2006), Op. cit., 116.
8. S. Ricketson, *The Paris Convention for the Protection of Industrial Property: A Commentary* (Oxford: Oxford University Press, 2015): 26–29.
9. S. Pollard, *Peaceful Conquest: The Industrialization of Europe 1760–1970* (Oxford: Oxford University Press, 1981): 271.
10. Ricketson (2015), Op. cit., 30–31.
11. There is a report of Vienna Congress of 1873 by the delegate of the British government T. Webster, *Congrès International des Brevets d'Invention tenu à l'Exposition Universelle de Vienne en 1873* (Paris: Marchal, Billard et Cie, 1877).

12. Y. Plasseraud and F. Savignon, *Paris 1883: Genèse Du Droit Unioniste Des Brevets* (Paris: Litec, 1983): 155–74; E. T. Penrose, *The Economics of the International Patent System* (Baltimore: Johns Hopkins Press, 1951): 48–55.

13. L. M. de Larra, *La Unión Internacional para la Protección de la Propiedad Industrial* (Madrid, 1887); T. Merly, *La Unión Internacional: análisis de la misma* (Madrid: 1890).

14. Plasseraud and Savignon (1983), Op. cit., 175–220; See also http://www.wipo.int/treaties/en/ip/paris/

15. *Conférence Internationale pour la Protection de la Propriété Industrielle* (Paris: Ministère des Affaires Étrangères, 1883): 21.

16. Penrose (1951), Op. cit., Chapter 4; Plasseraud and Savignon (1982), Op. cit., 205–9.

17. Ricketson (2015), Op. cit., 69.

18. See, for example, the article by the mining engineer and inventor Enrique Hauser: E. Hauser, *Bases fundamentales para una ley universal sobre patentes de invención* (Madrid: Teodoro y Alonso, 1900).

19. Royal Decree of 21/08/1884. See also *Industria e Invenciones* 189 (23/10/1886).

20. Ricketson (2015), Op. cit., 67–72.

21. Cited in A. Bogsch, *The First Hundred Years of the Madrid Agreement Concerning the International Registration of Marks* (Geneva, WIPO, 1991): 30.

22. BOPI No. 87, (01/04/1890); and *Procès-verbaux de la Conférence de Madrid de 1890 de l'Union pour la protection de la propriété industrielle, suivis des actes signés en 1891 et ratifiés en 1892* (Berne: Jent et Reinert, 1890).

23. See also Official Reports of Parliamentary Debates: *Industrial Conference at Madrid* (HC Deb 13 March 1890, vol. 342 c.717) and *The Times* (03/05/1890).

24. 'Conferencia Internacional para la protección de la propiedad industrial', *El País* (07/04/1890) and 09/04/1890. For a summary of press releases and editorials from newspapers of the time see: J. D. Montero Rius, *De Madrid a Madrid, cien años de marcas internacionales. Arreglo de Madrid 1891–1991* (Madrid: OEPM, 1992).

25. 'La Conferencia de Madrid', *El Imparcial* (02/04/1890).

26. See references in endnote 4 in this chapter. See also Chapter 1 of this book for additional references and a discussion on the dynamics of technology transfer to Spain during the nineteenth century.

27. Pretel (2017), Op. cit.
28. See Notes on Sources (appendix).
29. See R. Fox and A. Guagnini, *Laboratories, Workshops, and Sites: Concepts and Practices of Research in Industrial Europe, 1800–1914* (Berkeley: University of California Press, 1999): 158 and I. Inkster, *Science and Technology in History* (Basingstoke: Macmillan Education, 1991): 158.
30. AHOEPM, Privilegio 1212.
31. AHOEPM, Privilegio 1199.
32. For the series of agreements, the Ybarra Brothers signed with the company Bessemer & Longsdon and with the inventor Adrien Chenot, see P. Díaz, *Los Ybarra. Una dinastía de empresarios, 1801–2001* (Madrid: Marcial Pons Historia, 2002): 83–95. For the Bessemer system in Spain, see H. Bessemer, *An Autobiography* (London: 1905) and J. G. H., 'El sistema Bessemer en España. Su historia y su porvenir', *Revista Minera* XXX (1879): 353–354.
33. AHOEPM, Privilegio 1510.
34. For Henry Bessemer's profuse use of patenting for protecting his inventions see I. Inkster, *Science and Technology in History: An Approach to Industrial Development* (Basingstoke: Macmillan Education, 1991): 161 and G. Tweedale, 'Bessemer, Sir Henry (1813–1898)', *Oxford Dictionary of National Biography* (Oxford: Oxford University Press, 2004).
35. AHOEPM, Privilegio 1482.
36. C. K. Hyde, 'Iron and Steel Technologies moving between Europe and the United States', in D.J. Jeremy (ed.), *International Technology Transfer. Europe, Japan and the USA, 1700–1914* (Aldershot: Edward Elgar, 1991): 51–73.
37. From 1890 the Bessemer converter would be gradually replaced by open-hearth steel methods (e.g. Martin-Siemens), which yielded steel of superior quality and strength.
38. E. Fernández de Pinedo and R. Uriarte, 'British Technology and Spanish Iron Making During the Nineteenth Century', in C. Evans and G. Rydén (eds.), *The Industrial Revolution in Iron* (Aldershot: Ashgate, 2005):151–172.
39. H. Bessemer, *An Autobiography* (London: Offices of Engineering, 1905): 63.
40. *The Engineer* 256 (March 1898).
41. P. Sáiz and D. Pretel, 'Why Did Multinationals Patent in Spain? Several Historical Inquiries', in P-Y. Donzé and S. Nishimura (eds.), *Organizing Global Technology Flows: Institutions, Actors, and Processes* (New York:

Routledge, 2013): 39–59; For contemporary accounts on corporate patenting see 'Las patentes como base de los grandes negocios', *Industria e Invenciones*, No. 2, (13/01/1906): 13; and the discussion at the CIPA compiled with the title 'Grants of Patents to Companies', *TCIPA*, Vol. IX, (Session 1890–91): 126–8.

42. See endnote 29 in this chapter and C. Freeman, *The Economics of Industrial Innovation* (Cambridge: MIT Press, 1986): 176–178.

43. J. M. Ortiz-Villajos, 'Patents, What for? The Case of Crossley Brothers and the Introduction of the Gas Engine into Spain, c. 1870–1914', *Business History* 56 (4), (2004): 650–676; Sáiz and Pretel (2013), Op. cit.; P. Sáiz, 'Patents as Corporate Tools: Babcock & Wilcox's Business and Innovation Strategies in Spain', *Entreprises et Histoire* 82 (2016): 64–88.

44. A. Lozano, 'A Source of Modest Comfort: Las inversiones de Vickers en España, 1897–1936', *Historia Industrial* 16 (1999): 69–90.

45. TAEP    [D8941ABD],    28/10/1889;    [CE94114],    10/04/1894; [D9823AAH], 14/4/1898.

46. TAEP [CE88052], 02/04/1888.

47. TAEP [CE87044 and CE85003].

48. See F. L. Dyer and T. C. Martin, *Edison. His Life and Inventions* (New York: Harper & Bros., 1910): 971 and TAEP [D9226AAD], 14/07/1892.

49. D. Pretel, 'The Global Rise of Patent Expertise During the Late 19th Century', *Cambridge Working Papers in Economic and Social History*, 31 (2017).

50. T. Merly, 'La Unión Internacional para la protección de la Propiedad Industrial: análisis de la misma', *El Liberal* (20-31/03/1890).

51. In the late nineteenth century, some patent agents demanded that the state relax its requirements regarding the translation of patent and trademark documentation. See for instance the official request by the agent José Gómez Acebo to the State Minister. Royal Decree of 26/02/1897.

52. TAEP, Instructions to Agents, [D9823AAB], 03/10/1898.

53. TAEP    [D9823AAK],    (05/06/1898);    [D9823AAC],    03/18/1898; [D9823AAR], 3/6/1898.

54. TAEP [D9823AAG], 13/4/1898; [D9823AAB, TAEM 137: 321], 3/10/1898.

55. L. Galambos, 'The Role of Professionals in the Chandler Paradigm', in W. Lazonik, and D. Teece (eds.), *Management Innovation: Essays in the Spirit of Alfred Chandler* (New York: Oxford University Press, 2012): 125–146.

56. For the patent file, specifications, contract of assignment and notarised working of Marconi's patent on wireless technology in Spain see Patent no. 20041 (AHOPEM).

57. J. Johnson and J. H. Johnson, *The Patentee's Manual* (London: Longmans, Green, and Co., 1890), 351.

58. *Annuaires de L'Association Internationale pour la Protection de la Propriété Industrielle* (published from 1897); 'Asociación internacional para la protección de la propiedad industrial', *Industria e Invenciones* (02/05/1898): 205; Ricketson (2015), Op. cit., 75–6; M. Georgii, 'International Association for the Protection of Industrial Property', *The Inventive Age* 3 (March 1898): 42–3.

59. For some of these condemnations of the defects of the Spanish patent law of 1878 by Spanish agents see 'Patent Practice in Spain', *TCIPA*, Vol. VI (1888): 97–98; H. Graham, *On the Progress and Work of the Institute of Patent Agents* (London: Spottiswoode & Co., 1890): 18; For the interpretation of the Spanish patent law by Spanish agents see the correspondence of Gerónimo Bolibar and Alberto Clarke to the Institute of Patents Agents in 1901. Partially reproduced in *TCIPA*, Vol. XX (1901–2): 18 and 126–9.

60. 'Proposed Patent Law in Spain', *TCIPA,* Vol. II (1888–9): 33–4.

61. Letter from Julio Vizcarrondo to the Chartered Institute of Patent Agents (CIPA) read at the 28th ordinary meeting of this institute (London, 05/05/1886). A summary of its content is in "Spanish Patent Laws", *TCIPA*, Vol. IV (1885–6): 238–239.

62. Historical Dossier, Elzaburu Industrial Property Agency (Madrid, 2009).

63. Elzaburu Agency Private Records and the original powers of attorney kept in the patent documentation of the AHOEPM (see appendix on sources).

64. In the *International Directory of Patent Agents* for 1897 (London: William Reeves) Francisco Elzaburu advertised himself as 'Patent and Trademark business in Spain, Portugal and the Spanish-American countries. Careful work, Moderate Charges. First Class References'.

65. *Revista Ilustrada de la Banca, Ferrocarriles y Seguros* (25/11/1905): 550

66. AHOEPM Pat. no. 5602 and Elzaburu Agency Private Records, Registration Book for the year 1886.

67. J. S. Salaman, *Trade Marks: Their Registration and Protection in the United Kingdom and Abroad* (London: Shaw and Son, 1876).

68. For the high fees this agency charged at the turn of the century see the letter sent by Francisco Elzaburu in 1902 to a correspondent agent explaining the fees of the agency. Partially transcribed in *Journal of the Society of Patent Agents*, Vol. III, no. 27 (March 1902).

# 5

# The Colonial Dimension

**Abstract** This chapter studies the colonial dimension of the Spanish patent system during the nineteenth century. The first section explores the history of colonial patent institutions in the various Atlantic empires, with particular attention to the case of Latin America. The core of the chapter examines the regulation, administrative practices and technological culture of the imperfect patent system operating in Puerto Rico, Cuba and the Philippines between the 1820s and the 1860s. During the mid-nineteenth century, this neo-mercantilist institution served more as a system of technological information than as a means of protecting the rights of inventors, thereby reflecting the collective interests of colonial agrarian elites. The final part of the chapter traces the institutional reorganisation of the colonial patent system during the late nineteenth century, in a context of multilateral agreements and growing US influence.

**Keywords** Colonial patents • Technological culture • Institutional imperfection • Collective innovation • Plantation agriculture • Caribbean

© The Author(s) 2018                                                    **115**
D. Pretel, *Institutionalising Patents in Nineteenth-Century Spain*, Palgrave Studies in Economic History, https://doi.org/10.1007/978-3-319-96298-6_5

Colonial systems are likely the most telling example of the international diversity among patent cultures that prevailed during the industrialising nineteenth century. Imperial centres' interventions into the regulation of intellectual property rights in the colonies varied greatly through time and place. In general terms, the nineteenth century was a time when imperial powers saw such colonial institutions as being of little relevance, leaving colonial governments to pragmatically accommodate patent rights to their interests. In contrast to more formalised patent institutions, colonial systems were usually arbitrary and afforded weaker protection. The late nineteenth-century tendency towards the homogenisation of patent laws often prompted rejection from colonial authorities and agricultural landowners in overseas territories, who considered these laws of little use or preferred the more discretional local regulations. Institutional imperfection suited the colonies. Colonial authorities presumed that the economic gain engendered by technological change emerged primarily from innovation and not from actual inventive activity. This reinterpretation of the meaning and practice of patent protection was a comparative advantage for colonies insofar as it allowed, for instance, the unrestricted use of foreign technologies.

## A Diversity of Colonial Systems

Patents, taking different forms, had been awarded in several Atlantic colonies from early on. In the British case, North American colonies had granted patent monopolies discretionally at least since the beginning of the seventeenth century.[1] Massachusetts, Connecticut, Rhode Island, New York, Virginia, South Carolina and Plymouth all granted patents for inventions, especially in agricultural techniques, as reflected in the catalogue of patents prepared by the prestigious London lawyer Bennet Woodcroft.[2] Between 1839 and 1846 patents were also awarded in Texas.[3] There even was a Confederate Patent Office that granted 274 patents during the four years (1861–1865) it was in operation.[4]

Several colonies of the British empire, including India, Canada, Australia and several islands in the Atlantic and Pacific, had patent systems during the nineteenth century. For instance Lower and Upper Canada had separate patent laws from 1823 and 1826, respectively.

Following the principles of customary law (or common law) of Anglo-Saxon tradition, colonial patent systems in the British Empire functioned without metropolitan control until at least the English intellectual property law of 1852.[5] From that moment on, patents granted in England had to be recognised in the British colonies, thus restraining the free exploitation of many agricultural techniques. Still, the law of 1852 did not directly impact the regulation and functioning of colonial systems. By contrast, in the case of India, its colonial patent law of 1856 discriminated against Indian patentees vis-à-vis British patentees, excluding Indian inventors from the system.[6]

In 1864, 17 colonial laws remained in the British empire. Among them, the Australian example is particularly interesting.[7] Before the establishment of the Commonwealth Patent Office in 1904, each Australian colony managed its own independent patent system without adopting a uniform policy. Between 1852 and 1876, Queensland, Western Australia, New South Wales, Victoria, Tasmania and South Australia institutionalised patent systems without recognition of property rights among them. In practice, each system had different application procedures, registration costs, patentability standards, duration of monopoly and publicity. Australian patent policy promoted technology transfer and protected new industrial endeavours. Other British territories such as Hong Kong would not have enjoyed any patent legislation during the nineteenth century. Beginning in the 1870s, the British colonies would become more receptive to the assimilation of English patent legislation due to the opportunities presented by international agreements.[8]

In the French case, since the mid-eighteenth century, various Caribbean territories, such as Martinique, Guadeloupe, Reunion and Guyana, had established patent regimes that operated in parallel with the patent system in the metropole. In these colonies, the governors-general granted provisional patents, usually linked to the sugar industry, with the aim of stimulating the colonial economies. It would be between 1844 and 1849 that France regulated the conferring of patents in the colonies.[9] Until the end of the nineteenth century, the French system would continue to promote invention through monetary prizes, medals, subsidies, patent purchases, loans for the installation of machinery, technical translations and constraints to technological monopolies in strategic sectors.

There were several concessions of patents in Spanish colonial America well before the nineteenth century. From an early period, 'privileges of invention' were awarded to improvements in mining, refining, and agricultural techniques.[10] It seems that these privileges were the result of arbitrary concessions and loosely defined property rights that did not offer adequate legal protection to inventors.[11] However, the exact role of the various colonial governing bodies in the creation, codification and circulation of useful and reliable technical knowledge is a key question that has yet to be studied in detail. It was during the first third of the nineteenth century that the newly independent Latin American republics, including Mexico (1832), Brazil (1830) and Chile (1840), introduced modern intellectual property laws. In the second half of the nineteenth century, other nations in the Americas, such as Uruguay (1853), Argentina (1864), Peru (1869) and Venezuela (1878), would likewise establish intellectual property laws. By 1873, eight Latin American countries had patent systems. This contrasts with the situation in Asia and Africa, where only one country on each continent had patent regulations. This divergence can be explained, to a large extent, by the relatively later decolonisation process in Africa and Asia. In the year 1900, 15 Latin American nations had patent legislation, in contrast to four African nations and only two Asian.[12]

In Latin America, the newly independent nations attempted to overcome their colonial legacies and establish economic institutions that, like patent systems, supported the liberal construction of the state. The lawmakers of these new republics aimed to terminate the concept of patent rights as privileges of discretional absolutist prerogative.[13] Latin American republics adopted the French patent system as their preferred model, establishing patent laws that would promote the economic progress of the new nations. There were exceptions to this trend, such as Haiti and the Dominican Republic, which did not have patent laws during the nineteenth century, although they did adhere to international conventions on patent rights during the last third of that century. Some countries, such as Chile, Peru, Argentina and Brazil, granted 'patents of introduction', also known as 'patents of importation', a type of protection also accepted in Spain and its overseas territories. The scope of this legislation was to support the broad circulation of practical knowledge and the imitation of foreign technologies, regardless of whether the patent applicant was the original inventor.

During the last third of the nineteenth century, there was a gradual trans-formation of colonial patent systems. Among the countries adhering to the International Convention for the Protection of Industrial Property signed in Paris in 1883, were colonial powers such as Spain, Belgium, Portugal and France. Latin American countries, such as Brazil, El Salvador and Guatemala, would also be represented. Other imperial centres would join soon after that, such as Britain in 1884 and the United States in 1887. At the Paris Convention of 1883, attending countries addressed the regulation of colo-nial patent laws, although no agreements were reached. In the long term, the Paris Convention would have notable effects in colonial spaces.[14] During the 1880s and 1890s, the revision conferences that followed the Paris Convention incorporated direct references to colonial regulation, with few legal consequences due to opposition by Spain and the United States.[15]

At the First Revision Conference of the Paris Treaty, held in Rome in 1886, participants agreed that it would be the prerogative of the imperial centres to indicate which of its territories, colonies or possessions would become part of the Paris Union. On that occasion, the Spanish representa-tive was Mariano de Larra, director of the *Boletín Oficial de la Propiedad Industrial* (BOPI). Larra declared to the conference assembly that, per the instructions of the Spanish government, the islands of Cuba, Puerto Rico and the Philippines, as part of Spain, would have to be considered part of the Union.[16] Other colonies and dependent territories likewise became adher-ents of the Paris Union in the coming years, for instance, Syria and Algeria as part of France, Azores and Madeira as part of Portugal and Surinam and Curacao in the Dutch case. At the Fourth Revision Conference, celebrated in Washington in 1911, it was eventually decided that contracting imperial countries had the right to require the accession of all their colonies, posses-sions, dependent territories and protectorates to the Paris Convention.

During that same decade of the 1880s, and in response to the Paris Union, an inter-American patent system was formed, linked to the Pan-American Conferences held at the decade's end.[17] In the first Pan-American Conference, held in Washington between October 1889 and April 1890, the United States and many Latin American countries (including Mexico and Brazil) discussed the continental convergence on patent law. However, Santo Domingo, Cuba and Puerto Rico were not represented at this Conference; the first refused the invitation on the basis of its territorial disputes with the United States. The other two Caribbean islands did not attend because they were still colonies of Spain. At this conference, it was recommended that participants adhere to the treaty on patents that several American nations had signed at the South American Congress of Private International Law held in Montevideo some months earlier (August 1888 to February 1889). In 1902, in the second Pan-American conference held in Mexico, an agreement on patents, industrial drawings and trademarks was signed by, among other countries, Argentina, Bolivia, Chile, Mexico, Peru and Uruguay. The United States and Cuba, the latter newly independent from Spain, also adhered to the treaty.[18]

## The Virtues of Institutional Imperfection

After the Spanish-American independence processes, which took place between 1809 and 1826, Spain retained the colonies of its 'second empire' through 1898: Cuba, the Philippines and Puerto Rico (Fig. 5.1). These territories, especially Cuba, would become an essential part of the market and state structure of liberal Spain, both from a strictly economic point of view and in the political imaginary. The institutional and legal founda-

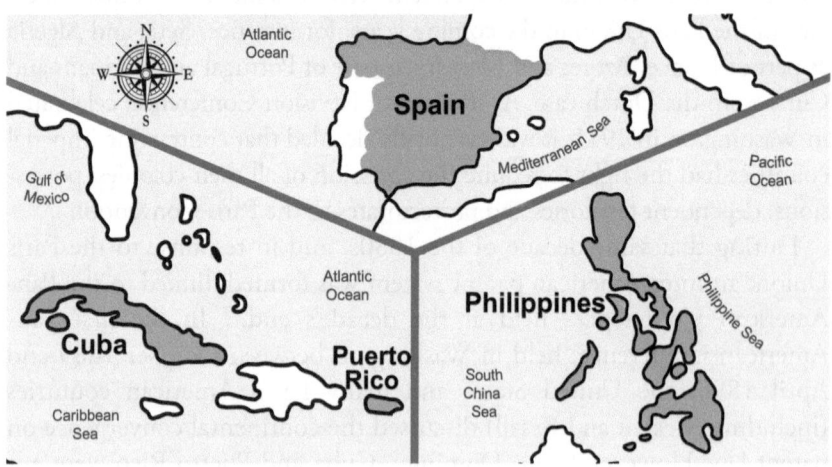

Fig. 5.1  Spain and its colonies in the nineteenth century

tion of Spain's modern empire was dual in nature. Despite the wealth of colonial economies and the consideration of these domains as a central part of the nation, creole elites were not involved in the broader political activity of imperial Spain, particularly the liberal political reforms carried out in the middle decades of the nineteenth century.[19] Beyond tax collection, the central government did not seem to have any particular interest in getting involved in policies to stimulate agricultural production and commercial activity in the colonies. Creole elites and local authorities, therefore, had enjoyed some autonomy in the administration of colonial economies and institutions since the end of the eighteenth century.

Following the Napoleonic tradition of dual metropolitan-colonial legal systems, Spanish nineteenth-century colonies had special colony-specific laws. Regarding intellectual property rights, there were particular regulations for the colonies (Royal Decree of 1833 and Royal Decree of 1880) that differed from the laws that were applied in the Peninsula (laws of 1820, 1826 and 1878).[20] The colonies and the metropole had, during the years in which the colonial patent system (1833–1898) operated, separate patent registration offices and official publications.

The administrative communication between the two systems was irregular at best.[21] The colonial law of 1833 was not officially publicised in metropolitan Spain until 1849.[22] Hundreds of the patents granted in these three colonies were, thus, directly administered within these territories themselves; technical information relating to colonial patents was not regularly communicated to the central administration in Madrid. Under this dual legal infrastructure the colonial patent system became an informal institution, characterised by neo-mercantilist practices and controlled by colonial socio-political elites and their corporations. In practice, the granting of patents in the Spanish colonies functioned as a system of privileges in which technological monopolies were granted only after an examination of the utility of the invention for colonial economies and local interests.

Between the 1830s and the 1860s, colonial patents were granted in ad hoc discretionary and arbitrary case-specific decisions. Ambiguous and exceptional legislation gave the colonial elites the authority to grant monopolistic privileges. Local colonial authorities assumed the power of issuing grants conferring patents within their territory following a utilitarian scheme whose objective was to promote colonial agrarian economies that produced valuable commodities for global markets. The colonial law of 1833 explicitly restricted the granting of patents of introduction related to innovations in agricultural techniques, particularly those related to sugar inventions, leaving the final decision up to colonial authorities. As in the metropolitan system, a much shorter protection period was granted to non-original inventions (five years) than to original inventions (5, 10 or 15 years, at the wish of the patentee). The already advanced state of Cuban plantation agriculture, namely its sugarcane sector, was the reason behind this exceptional provision in the granting of colonial patents. Cuban planters and their various institutions were already well aware of the cutting-edge machinery, instruments, devices, procedures and scientific methods that were ubiquitous in this industry.

The patentability criteria were also pragmatic, which in practice restricted the granting of patents in strategic or advanced economic sectors. The institutional imperfections of the colonial patent system offered clear advantages. Foreign inventions were introduced without the additional costs that a strict enforcement would have entailed. Colonial authorities pragmatically administered the system, guided by the changing imperatives of their economies. The authorities understood patent rights as exclusive privileges that enabled inventors to pursue economic activity, rather than as a natural or moral right. The contract between the patentee and the colonial administration centred on the actual functioning of the new technology in a given territory. Colonial governments and elites were primarily interested in the potential agro-industrial and commercial gains of introducing new foreign inventions at a low cost.

The administrative practices of the metropolitan and colonial systems were quite different. The colonial system lacked any sort of professionalised or specialised patent administration. The Spanish office, meanwhile, had—with variations throughout the nineteenth century—some personnel dedicated to the routine administration of patents, although lacked the modicum of economic resources. In the colonies, employees of political organisations or colonial corporations were left in charge of the registration, evaluation and advertising of patents. The administrative practices were not standardised, nor was a bureaucratic expert body in place to train examiners and officials in the granting of patents. The law left the final decision about the granting of patents to the discretion of colonial political authorities. This meant that each of the colonies had its own patent office, which in practice was run by colonial governing bodies, urban societies and landlords' corporations. In the case of Cuba, its distinctive patent institution was administered by the Havana City Council, the Real Junta de Fomento (Royal Development Board) and the Sociedad Económica (Economic Society).[23] Following the law of 1833, concessions of privileges were to be published in the respective colonial *Diarios de Gobierno* and *Gaceta de Madrid*.

In the colonies, there was no patent office dedicated solely to the administration, concession and publicity of patents. Instead there was a council in charge of these and other many issues. The main body in charge of the processing and registration of patents in the case of Cuba was the Real Junta de Fomento, a powerful corporation founded by the Havana oligarchy.[24] Between the 1830s and 1860s, this institution's administrative patenting requirements were inconsistent; it frequently granted privileges without the applicant having furnished technical specifications, detailed claims or a drawing of the invention. Final decisions on concessions were the prerogative of political and economic creole elites, who were principally interested in modernising sugar plantations and refineries. The invention was considered novel when it was not public knowledge on the island despite the fact that it could be known abroad.

The accomplishment of such official examinations of utility and novelty required some specialised knowledge and practical expertise on the part of those conferring the patents. For final decisions on the granting of patents in Cuba, the Junta de Fomento commissioned reports from the Sociedad Económica, the Havana City Council and independent technical experts such as the Catalan inventor Mariano Vieta, a pharmacist, or chemist José Luis Casaseca, originally from Salamanca, Spain, but educated in Paris. For example, on 18 February 1850, Casaseca wrote a report supporting the concession of a patent for a technique for the purification and clarification of sugarcane requested by the landowners Joaquín and Pablo Arrieta.[25]

The reports of the colonial elites after examining inventions were, in the words of the members of the Junta, '(…) stricter than indulgent for the (privileges) of introduction and very cautious in those of improvement'.[26] Frequently, to prepare these reports, commissions were established to carry out inspections of the new procedures in plantations, factories, railways and workshops.[27] The preparation of these reports was facilitated by the close social proximity and interpersonal contact that existed between patentees and those registering and examining the applications. The lawyer and secretary of the Real Sociedad Económica, Rafael Matamoros, concluded in 1843 that 'before issuing certificates of privilege for inventions and introduction of machines and improvements for the promotion of agriculture, the Government has consulted the opinion of the Corps, and we can guarantee without fear of deception, that our reports have followed the best principles of economic science and the purest ones for the welfare of the island'.[28] This had not always been the case; in 1836 the Sociedad Económica itself denounced these reports as sometimes inaccurate and arbitrary.[29]

Despite the rhetoric of the 1833 law, the Spanish colonies did not develop a consistent system of granting patents. The social context in which colonial patent regulations were administered altered the way these exclusive grants were perceived and implemented. The ambiguous colonial legislation left room for unsystematic administrative practices in the awarding of patents. Each patent application was studied case by case through an ad hoc procedure. The primary variable that determined the granting of a temporary monopoly was the economic benefit of the invention for the colonial agrarian economies, rather than the intrinsic novelty of the invention itself or the rights of the inventor.[30] Unlike what happened in metropolitan Spain, where the patent institution operated as a registration system without technical examination, in the colonies there was an exhaustive examination of the utility of patented inventions for the colonial agrarian economies.

The incremental and experiential nature of many of the chemical and mechanical innovations introduced in colonial economies made it challenging to detect the unauthorised use of patent rights. In their decisions, apart from an assessment of an invention's novelty, colonial authorities considered the cost of the new machinery, the effects on colonial commercial life and alternative inventions already in use. In creating a system with an inconsistent examination of inventions, colonial authorities safeguarded the prevailing interests of ruling groups and the investments of colonial elites. Of course, in the end, their decisions were also constrained by imposed geographical, environmental and geopolitical conditions.

Metropolitan and colonial patent systems had different institutional evolutions. Divergent patent cultures were a consequence of different economic imperatives (related to the endowment-resource base) as well as locally prevailing colonial interests and ideologies. Until the 1860s, the barrier to patenting in the colonies was not only the prohibitive cost of applications—a consequence of additional registration fees and the need to use chains of international intermediaries—but especially the fact that the examinations were carried out by local political corporations. In contrast, in metropolitan Spain the barrier to patenting was primarily the cost of registration and not the examination of the patent; as was the case in England, Spain had a registration system that operated without any technical examinations before granting the patent. However, both the metropolitan and colonial systems had limited resources with which to ensure substantive compliance with the regulations.[31]

# A Collective System of Information

Colonial institutions such as the Juntas de Fomento used a variety of alternative strategies to reward inventors, many of which served as greater incentives to innovation than patents. A central defining feature of the Cuban technological system in the middle decades of the nineteenth century was that local governing corporations promoted innovation through prizes, subsidies, study commissions, exhibitions, public trials, expeditions abroad, the translation of manuals and commercial monopolies. These complementary incentives to innovation frequently clashed with the granting of patents insofar as they often took the form of 'privileges of invention', when in reality they did not meet, in any case, the legal requirements for this type of protection. Although many of the incentives were ancillary to patenting, they operated outside of colonial patent regulations.

The rise of the Cuban sugar industry was strongly associated with increased institutional facilities for practical knowledge generation, circulation and testing. Planters, scientists, lawyers, machinists and commercial houses organised an information system around colonial corporations. Knowledge sharing also took place at sugar factories and plantations, as well as through both the migration of skilled workers to Cuba and Cuban planters' travel abroad. Colonial governing societies created an infrastructure for information exchange whose objective was to avoid technological secrecy and favour the exchange of techniques, rights, experts, texts and ideas. In the period from 1820 to 1860 and possibly beyond, this model of promoting innovation thrived insofar as it encouraged the circulation of novel mechanical and chemical inventions in sugar refining, such as evaporation devices and clarifiers, as well as agricultural practices in the sugarcane yield, such as ratooning and the introduction of fertilisers. These technological inputs in the sugar industry were combined with larger infrastructural and transport transformations such as railroads and iron-steamships. The result was that this model of governance for technological innovation in the plantation complex—with its prevailing slave labour—brought the Cuban sugar industry to an unknown level of technification and engineering capacities with ever-increasing production levels. The on-site sugar processing at large Cuban *ingenios* in the western provinces of the island epitomised such a technological and organisational surge.

Meanwhile, elite landowners implemented a strategy to disseminate technological information and statistical data related to inventions. This information was transmitted to the public sphere in official diaries, corporations' publications, reports and monographic treatises. Among the publications worth mentioning are *Memorias de la Sociedad Económica*, the *Gaceta de la Habana* and, in the last third of the nineteenth century, *La Revista de Agricultura* (Journal of Agriculture) published by El Círculo de Hacendados, the Cuban association of landlords founded in Havana in 1878. Another publication of the day, short-lived but worthy of mention, was *Los Anales de Ciencias, Agricultura, Comercio y Artes* (Annals of Science, Agriculture, Commerce and Arts). Edited in Cuba by Spanish naturalist Ramón de la Sagra between 1821 and 1835, it disseminated information on sugar technologies, botany and scientific agriculture. Translations of foreign technical and agricultural treatises with the economic support of the Junta de Fomento were also frequent. A good example is a translation and publication in 1844 of Charles Derosne and Jean Francois Cail's treatise on the production of sugar in the colonies, that was distributed free of charge among Cuban sugar planters and public libraries.[32] In this publication, Derosne and Cail described their steam evaporation device, a vast and costly equipment that had already been introduced to Cuban sugarcane refineries a year earlier.

This deliberate programme of transmission of technological information and agricultural practices to the public domain was based on a collective conception of innovation, in which landowners freely released technical information resulting from practical experimentation in the plantation complex. Technological inventions and new cultivation methods, such as seed breeding and improvements in fertilisers, were largely conceived as non-exclusive public goods. Planters' corporations also established institutions for technical and agricultural learning and training. For example, the Sociedad Económica set up in 1845 a School of Machinery that specialised in technical drawing, mechanics, geometry and iron works. This school of mechanics, which operated out of the same building as the Sociedad Económica, had the declared purpose of reducing Cuban dependency on the foreign machinists who worked on sugar plantations, railways and steamships.[33]

Colonial elites and their corporations did not oppose patenting, rather they took a middle ground between protecting patentees and securing local interests. Patent rights were often granted, but their scope and enforcement were limited. In other words, the colonial patent system was intended more as an institutional infrastructure for the exchange of information than for claiming property rights. Competition among sugar planters played out not at the local level but in the international market, due to the rise of new producing regions and the increasing scale of production. It seems that competition among Cuban sugar-mill owners focused not only on pecuniary questions but also on social status and reputation, including innovations and technological accomplishments.[34] Rejection of patents in other sectors beyond the agricultural was also common. For example, in 1844 the Junta de Fomento dismissed the request of Juan Sparron to protect a new metal compound for making machines. According to colonial authorities, the request did not meet legal standards because the invention was already well known and in use in Havana's railway repair workshops.[35]

Colonial corporations acted as an emporium of information centred around the Junta de Fomento and the Sociedad Económica, which served as a collective space for the public promotion and dissemination of technological information. The Junta de Fomento established commissions to encourage the introduction of agricultural improvements and new technologies on the island, showcasing their placement, operation and harvest results. It also sponsored landowners, government officials and Cuban scientists to travel to North America, Europe and other Caribbean islands in search of practical agricultural knowledge, machinery, chemical inputs and experts. The new techniques described in the reports resulting from these commissions would not be eligible for patent protection until more than three years had passed since their awareness in Cuba—and then only if they had not been put into economic practice on the island.

For example, in 1848, the Junta de Fomento commissioned José María de la Torre, professor of Geography at the University of Havana and member of Sociedad Económica, to visit the United States for several months, including the Louisiana sugar plantations and the Washington Patent Museum.[36] The notes of this commissioner, which were published in the *Memorias de la Real Sociedad Económica* and in several Cuban newspapers, included references to the main North American cultivation and cattle-raising treaties of the time, as well as to official statistics and models of machines and inventions. De la Torre brought to Cuba seeds of different kinds, including cotton and wheat, as well as samples of products, small instruments and machinery. Among them, de la Torre presented in Cuba the sugar evaporation device of Norbert Rillieux, which had American and French patents. Other individuals—including scientists, planters and politicians—likewise received travel stipends from the Junta de Fomento to examine foreign technological inventions, resulting, for instance, in a one-year stay in Europe (including Belgium, England, France and Austria) by the prestigious chemist José Luis Casaseca in 1842; a five-month expedition by landowners Ramón Arozarena and Pedro Banduy to Jamaica in 1828; and a visit by Alejandro de Olivar to England and France in 1830.[37]

There is no better written manifestation of Cuban technological culture during the mid-nineteenth century than patent files. The extensive reports commissioned by the Junta the Fomento show that the technological culture of Cuban sugar refineries was marked by collective practical experimentation, professional expertise and incremental innovation. Patenting related to the improvement of sugar production in the middle decades of the nineteenth century was characterised by an effort to adapt foreign technologies to the island's particular tropical conditions. It was an innovation model of tacit knowledge deeply embedded in the microculture of the plantation complex, which contrasted with the more formal technological culture that could be found in the industrial enclaves of metropolitan Spain and its engineering schools (Fig. 5.2 illustrates the scope of the technification of sugarcane processing in Cuba during the mid-nineteenth century,

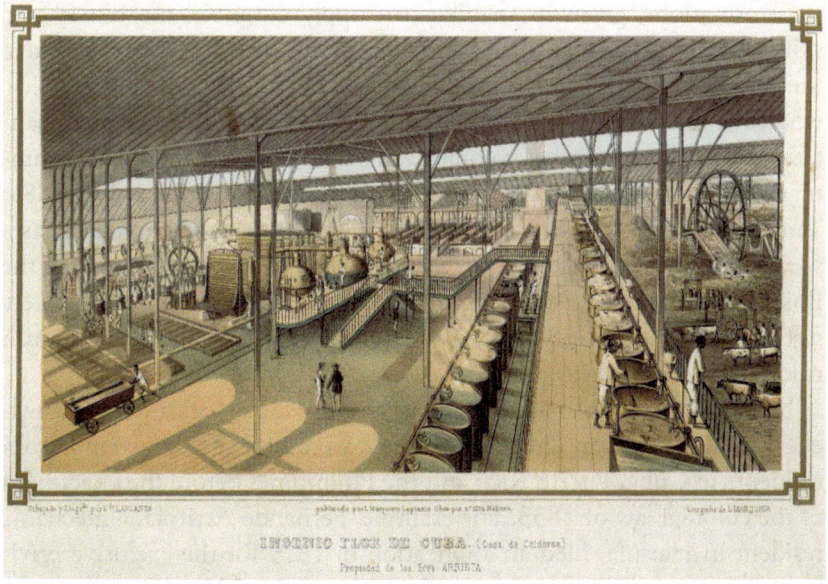

**Fig. 5.2** A sugar refinery in mid-nineteenth-century Cuba. (Source: Justo G. Cantero, *Los ingenios: colección de vistas de los principles ingenios de azúcar de la isla de Cuba* (Havana, 1857))

not to mention the coordination of slave labour, skilled expertise, animals, technology and agriculture in the plantation complex).[38]

## Colonial Patent Networks

Spanish Caribbean colonies experienced intense economic transformation during the central decades of the nineteenth century. The colonial elites in these territories vigorously promoted commodity production, railroads and commercial activities. Between 1820 and 1860, the Cuban and Puerto Rican sugar industries witnessed a process of technological modernisation, a transition towards a large-scale industrial model and intensive investment in machinery and processing equipment, facilitated by the presence of transnational qualified experts, along with the persistence of slave labour.[39] As shown by the trade balances and records of machinists, metropolitan Spain played a minor role in Cuba's technological transformations, given its inability to supply advanced machinery and skilled technical workers to the sugar mills. This situation contrasted with the situation on other sugar-producing islands, such as Java and Jamaica, where the metropolises were active in the process of mechanising the plantations of their respective colonies. This situation was eloquently summarised by the industrial engineer and patent agent José Alcover in 1884: 'our Antilles were chosen by certain foreign constructor companies as a special market in which to place what they call export machines'.[40] In the Philippines, large-scale sugar production and agrarian mechanisation were not adopted until the 1860s, but particularly after the collapse of Spanish rule at the turn of the century.[41]

Social networks built around colonial patent systems facilitated knowledge and information circulation. There were requests for patents from residents in Cuba, Puerto Rico and the Philippines before the enactment of the colonial law of 1833. For example, Fernando Arritola, a mechanic resident in Havana, filed an application in 1819 for the exclusive privilege of manufacturing a sugar alembic and, in 1820, protection for improvement of this same distiller.[42] The dossier presented by Arritola in 1819 lacked a design or drawing, but the Committee of Agriculture, Industry and Arts of the Spanish Parliament in Madrid agreed in 1820 to approve his application and determined that, given the distance of the overseas possessions, applications for privileges of invention from colonial residents could be presented directly to colonial authorities.

Similarly, in April 1824, Lorenzo Calvo and Domingo Rojas,[43] prominent businessmen from Manila with significant political influence, were granted a privilege of introduction for 12 years to install an English-made complete steam-operated iron works in the Philippines, intended to melt, laminate and refine iron.[44] In December of that same year, the Royal Conservatory of Arts in Madrid granted these same merchants another introduction privilege for 12 years for a spinning and twisting machine of foreign construction. Domingo de Rojas, one of the most prominent merchants in the Philippines, had interests in the tobacco trade, the iron industry and the cultivation of sugar and introduced machinery in these realms. Rojas enjoyed important positions in colonial institutions of government. Lorenzo Calvo was a major player in the Philippine economy and a promoter of iron mining in the islands. Rojas and Domingo were partners in various ventures.[45]

Between 1826 and 1898, some 2600 applications for patent of inventions and introduction were administered in Cuba.[46] The applications for protection concentrated on inventions related to the production of sugar, construction materials, fuel, transportation and machinery, but there were also applications for various improvements in organisational procedures. Several applications for improvements in plant cultivation methods and agricultural techniques were also submitted, as Spanish legislation of the nineteenth century had included them in the scope of patent protection. The number of patents on agricultural innovations in Cuba, many of them spinoffs from other sectors, was significantly larger than in metropolitan Spain. Moreover, between 1820 and 1898, 575 applications for patents of invention were submitted in the registries of metropolitan Spain for residents in Cuba, the Philippines and Puerto Rico (499 for Cuba, 61 for Puerto Rico and 15 for the Philippines). The great majority, 455 applications, were submitted between 1880 and 1898. Dozens of patent rights granted in metropolitan Spain were extended to the colonies through administrative procedures at Ministerio de Ultramar.[47]

The period with the highest number of applications was the 1850s, with over 800 requests, coinciding with the technological transformation of the island's sugar sector. The 1860s saw a dramatic decline in the number of applications registered in Cuba, with applications dropping to around 500. From that decade, the Cuban economy was waning due to political conflict, the end of slave labour and new sugar-refining technologies that increased international competition. In the decade of the 1870s there were only 130 patent applications submitted. The decline in applications contrasts with the patterns at the international level, where there was an escalation in patenting activity from the 1870s due to economic globalisation and increasing facilities for international patenting. This sharp reduction may also be attributable to an inconsistency in the historical documentation of patents. Indeed, due to the absence of adequate records and the unsystematic and ad-hoc granting procedures of the time, long-term quantitative analyses of Cuban patents are problematic.

Patent applications in Cuba during the central decades of the nineteenth century were related to both foreign and domestic inventions. There was a predominance of 'creole' inventions, a result of transnational collaborations between local and foreign actors. In fact, any clear distinction between international and domestic inventions was dissolved.

The plantation complex was a site for the development and circulation of transnational innovations, not a backward space relegated to importing technology. However, patent rights were weak and difficult to enforce and therefore of little use to plantation owners in the face of legal uncertainty. This made colonial patents a limited source of revenue for patentees, restricting, in turn, the market for intellectual property rights and the manufacturing of machinery and tools in the colonies.

Among the patentees in the Cuban system, there was the American Alfred Cruger (chief engineer of the railway line from Havana to Güines), the New York-based commercial house Drake Brothers and Company, the Bell Telephone company, the chemist José Luis Casaseca, the professional photographer Esteban Mestre Aulet and the plantation owners Juan Poey and Wanceslao de Villaurrutia. The majority of foreign inventions were requested or intermediated by residents of Cuba. Among these intermediaries, the planters and machinists were most prominent insofar as they not only processed the foreign requests but also adapted the new technologies of sugar production to the island's tropical conditions, like its climate and topography. For instance, French engineer Pierre Theodore Vaurigaud, a professor at Havana's School of Machinery, intermediated in 1851 to obtain a patent on behalf of Enrique Oliveiro Robiuson for a steam machine used in the production of sugar.[48]

The migration of transnational machinists, engineers and skilled workers increased the capacity of the Cuban sugar industry to assimilate new technologies. The presence of advanced technical experts also created new patterns of patenting and paths of knowledge transmission. Foreign machinists, chemists and engineers, as well as other skilled workers, were involved in the Cuban patent system, acting either as individual patentees or engaging in relationships with inventors, merchants and planters. Particularly active during the mid-nineteenth century were Fernando Klever, Ezra Dod, Hiran Havens, Charles Edmonstone, Elisha Fitgerald, Michael Glynn, Edward Beanes and James Ross. Many of these 'travelling' transnational engineering workers appeared in the registers of foreign machinists kept by colonial authorities in the sugar-making provinces of Santiago, Cárdenas, Matanzas and Cienfuegos.[49]

The large presence of foreign machinists in Cuba during the 1840s and 1850s was repeatedly highlighted by high-circulation American technological journals such as *Scientific American*. In 1855[50] this magazine observed that 'During the sugar cane season in Cuba, say from November to April, there are usually employed on the various plantations about twelve hundred machinists as engineers and repairers. Few of these machinists are Cubans, and few of them remain the whole year on the island. A large number are Scotsmen, a few English, while the United States furnish a large share'. Similarly, only a few years earlier, in 1851, this magazine had declared that Cuba was 'almost wholly supplied with machinist from the United States. There is in nearly every plantation in Cuba a sugar mill driven by steam engines, built usually in New York or Boston'.[51] Expatriate networks of foreign engineers working in Cuba and Puerto Rico would continue growing in the last third of the nineteenth century, particularly in three economic sectors: sugar production, mining and railways. Meanwhile, the rapid expansion of US trade and investment in the Spanish Caribbean further encouraged the emigration of American engineers.

The widespread movement of foreign engineers into the Spanish Caribbean had repercussions for transnational patenting activity. These foreign machinists residing in Cuba further registered the inventions they had conceived in the plantations of the Spanish Antilles in patent offices in Britain, France and, above all, the neighbouring United States—countries that offered superior security, legal guarantees and opportunities for obtaining economic profits from patents.[52] A good example of expert participation in Cuban plantations is the British engineer and inventor Edward Beanes, who had over two decades of experience working on Cuban plantations. In 1865 Beanes obtained a patent in the United States for improvements in the refining of sugarcane juice, an invention that he had also registered in Cuba and the United Kingdom.[53] Beanes had been an engineering consultant for the British engineering firms Mirless Watson and Fawcett Preston, acting as a technical intermediary and commercial agent.

Puerto Ricans also patented technologies in Cuba. For example, in 1868, Puerto Rico resident Carlos Federico Schomburg requested, through his representative in Cuba, José Peligero de Lama, two privileges, one for a new system of mounting cookware to raise sugar and a second for a system of irrigation by means of tubular wells in combination of pumps. Both were granted.[54] Another Puerto Rican plantation owner, Juan Ramos, patented different procedures for the purification and clarification of sugarcane in Puerto Rico, Cuba, metropolitan Spain and the United States.[55] Ramos sold some of his patent rights to Cuban and US entrepreneurs. A singular case involved a machine to grind sugarcane whose patent rights Juan Ramos agreed to transfer to Julio Leneven and Jose Coste. Ramos sold his rights with the written condition that the new owners fix a maximum price of 425 pesos for the sale of this device in Cuba, of which 25 pesos had to be donated to a charity house in Havana.[56]

Business partners Charles Derosne, a chemist, and Jean-François Cail, a manufacturer, were especially active patentees in the Cuban system from the 1830s. Their inventions were a worldwide commercial success, although some of their numerous patents were deemed invalid by smaller independent inventors such as Howard and DeGrand.[57] In 1842, Derosne and Cail demanded that Cuban authorities grant them a privilege of invention for a modern sugar evaporation device that adapted to sugarcane the advances already made in the production of sugar beets.[58] The colonial corporations of Cuba acknowledged the exceptional progress that this technical installation contributed to the sugar industry. However, the Junta de Fomento decided not to grant a patent after collecting three detailed reports, and instead awarded the Frenchmen a monetary prize for their invention. The reason given by the Junta de Fomento was that this invention had already been introduced in Cuba and that several Cuban institutions had provided funds for its broad diffusion in the island. The report of the Sociedad Económica established that Dersone and Cail were not the inventors of the main scientific base of the machinery and that all the parts that made up the device were manufactured in France and not Cuba. For these reasons the report of the Sociedad Económica concluded that the concession of the patent monopoly would be detrimental to Cuban commerce and particularly to the island's sugar producers. Similarly, in 1845 the company set up by these two entrepreneurs, Derosne and Cail, was denied a request for a privilege of introduction for a new method of purging and crystallising sugarcane. The commission formed for this purpose by the Junta de Fomento justified this particular rejection with the argument that technical improvements in sugar production were already quite advanced on the island. The report made clear that exceptions in the granting of patents were necessary to maintain the profitability of the sugar industry and the general diffusion of foreign technical innovations.[59] Other requests for protection were granted to this French company by the Junta de Fomento, as, for example, in the year 1850 for a machine designed to produce sugar without employing animal charcoal.[60]

# The Colonial System in Times of Convergence and War

During the two last decades of the nineteenth century, there was a reconfiguration of the patent system in the Spanish colonies in a context of multilateral agreements and expanding US influence. Despite the new international arrangements on intellectual property rights of the 1880s and 1980s, a special colonial patent institution would continue to operate in the islands of Cuba, Puerto Rico and the Philippines until 1898. It was through the patent law of 1880 that the Spanish government had reaffirmed the exceptional character of the self-governed colonial system. The governors-general of the colonies conserved the prerogative of granting patents with effect only in the overseas provinces.[61] As contemporary legal experts such as Cuban lawyer and public notary Francisco García Garófalo admitted,[62] two different patent systems continued to coexist during the late Spanish empire: one, legislated for the metropolis, in which applications had effects in all Spanish domains, and another, colonial, in which patents were administered by overseas authorities.

Unlike the legislation of 1833, the law of 1880 established that patents granted in mainland Spain had legal effect for patentees in all the Spanish domains, without requiring them to pay four separate registration fees. This meant in practice that every metropolitan patent right was considered granted not only on the metropole but also in the overseas provinces. Registration fees in the colonies, following the same principles of the patent law of 1878, would become progressive. These legal changes favoured a steady increase, between 1880 and 1898, in the number of patent applications arriving at metropolitan offices from residents in the colonies. However, although the patents issued in Spain were valid in all the colonial domains, a specific request to the Ministerio de Ultramar and a legalised testimony were required. Patents could also be legalised directly in each colony, as overseas authorities kept a register of patents issued and published the information of the concessions in official colonial gazettes. The colonial patent system in the later nineteenth century lent itself to abuses and legal insecurity and was denounced, among others, by industrial engineer Gumersindo Vicuña, professor of mathematical physics at the Central University of Madrid and former Director General of Agriculture, Industry and Commerce.[63]

According to Vicuña, the special law not only introduced a metropolis-colony dualism but also was responsible for the many variations in the administration of patents among the colonies.

After the reform of 1880, any patent granted overseas could as well be extended to other Spanish domains, including the metropole, through a request to the Governor General, at no additional cost. The difficulties encountered by foreign applicants to satisfy the twisted legal and administrative requirements of the 1878 Spanish patent law, which was extended to Cuba in 1880, led to a proliferation of colonial intermediary agents. It seems, however, that the mushrooming of colonial intermediaries was already commonplace before the 1880 law. For instance, in 1877 an editorial in the Madrid mechanic journal *La Crónica de la Industria* (Industrial Chronicle) had complained that '(…) if you want as a Spaniard to protect your invention in Spain as a whole, then it is necessary to cross the seas, name agents in Cuba, Puerto Rico and the Philippines, get three new patents, and spend practically a fortune'.[64]

Professional patent agencies became central in colonial patenting after the legal reform of 1880. Three patent agencies dominated the lucrative activity of patent intermediation in the Spanish colonies: Elzaburu (founded by the Puerto Rican lawyer Julio Vizcarrondo); the law firm Clarke, Modet & Company and the consulting engineering firm Centro Auxiliar de la Industria. Big multinational companies producing and selling machinery, such as the British Duncan Stewart and the French Fives Lille, hired these intermediary agencies to issue Spanish colonial patents during the 1880s and 1890s. Agents were often required to arrange the compulsory working of inventions in colonial sugar plantations, where European and American engineering companies had some of their largest markets. Engineers working for foreign multinational corporations similarly acted as colonial agents in the management of patents, such as Adrian Dumoulin, a French civil engineer and patentee himself who worked for the firm Fives Lille, among others.

Apart from a special patent regulation, the colonies would enjoy special legislation on trademarks, industrial designs and models from the 1880s. These additional colonial laws on intellectual property rights were introduced in Cuba and Puerto Rico by the initiative of Manuel Aguirre de Tejada, Minister of Ultramar, immediately after the agreements of the Paris Convention of 1883.[65] The Spanish government justified the existence of distinct colonial trademark laws by citing the need for exceptional protection of the tobacco industry in the two Antilles. In this regard, the renowned patent attorney José Pellá y Forgás, author in 1892 of *Las Patentes de Invención y los Derechos del Inventor* (The Patents of Invention and the Rights of Inventor), one of the most extensive and outstanding books on the Spanish patent law of the time, wrote that 'the anomaly of having legislation on industrial models for overseas and not having it for Spain….can only be explained by the laziness that characterises Spanish governments'.[66] In reality, even the colonial elites did not seem comfortable with the gradual national and international homologation of the colonial system of industrial property. In this sense, Francisco Zayas, director of *Revista de Agricultura*, the official organ of Cuban planters association El Círculo de Hacendados, declared in 1880 that it would be better for the colonial economy if the new machines and apparatuses employed in its sugar industry were not protected by patents of invention that constrained innovation.[67]

In the later nineteenth century, there was growing interest among American industrialists and companies in the economic opportunities in the Spanish colonies. A good example of this trend was the numerous articles that appeared in the industrial export magazine *La América Científica e Industrial* (Industrial and Scientific America), published monthly in New York by the patent agency Munn & Company between 1890 and 1905. This 'patent magazine' was accompanied by the *Scientific American Export Edition* and was primarily aimed at Spanish-speaking Latin American countries, where the patent agency Munn & Co could intermediate among systems and facilitate machinery trade. Cuba and Puerto Rico were the subjects of frequent articles in this journal, many of them explaining the opportunities that the sugar industry and railways presented to inventors, engineers and machinery manufacturers in the United States. The superiority of American technology and the industrial exhibitions held in the United States, such as the Chicago Columbian Exposition of 1893, were feature contents of this journal. American machinery manufacturers also advertised their tools and equipment in every issue of this publication. The generalist Spanish press from 1895, on the contrary, saw this growing US technological superiority as a danger for Spanish interests in Cuba and a potential contributing factor to the island's move towards political independence.[68]

In the last third of the nineteenth century, the number of applications for patents in Cuba dramatically declined. At the same time, there was a relative increase in the number of applications by American inventors and companies. For example, the New York division of Babcock Wilcock & Co. applied in 1885 for a 10-year patent for an improvement in the construction of a steam boiler.[69] The Junta de Fomento granted the patent. An extension on the period of compulsory working was requested some months later by Spanish engineer Alberto Verastegui, a representative of the company on the island. The Junta de Fomento accepted the request as the machinery protected was a large-scale and expensive technological equipment. Frederic Cook, an engineer working for the company Babcock Wilcox, also obtained a patent in Havana, a few years later, in 1889, for a complete automatic device to burn green bagasse in combination with a battery of sectional steam boilers. On this occasion, the duration of protection was 15 years. During the 1880s the British branch of Babcock and Wilcox also had important commercial interests in Cuba, where it sold expensive sugar processing machinery.[70]

During the 1880s Edison companies were especially active in colonial settings, including India, South Africa, Ceylon and Australia. As for the Spanish-American colonies, Edison set up several companies in New York to control, advertise and licence Edison's patented technologies in Cuba, Puerto Rico and 'other Spanish colonies': the Electric Light Company of Cuba and Porto Rico, the Edison Electric Light Company of Havana, the Edison Spanish and Colonial Electric Light Company and the Havana Electric Light Company.[71] These firms were incorporated mostly to control patents and commercialise technology, such as electric lighting plants for sugar factories and cities. From 1882 through 1884 the Edison Spanish Colonial Light Company even had an exhibition plant in Havana, with American Edward Beardsley as chief engineer.[72] According to electrical engineer Thomas C. Martin and Edison's patent lawyer Frank L. Dyer, Edison companies were granted 12 patents in Cuba.[73] In January and February of 1883, the Edison Spanish Colonial Light Company also obtained several patents in Madrid on electric production, distribution and incandescent lamps.[74] Soon after these colonial companies were established, Thomas Alva Edison assigned to them some of the patents he was granted in mainland Spain.[75]

In the aftermath of the Spanish-American War, the regulation of intellectual property rights in the colonies changed. Following decades of growing US economic influence, Cuba, Puerto Rico and the Philippines came into the geopolitical orbit of the United States. The increasing US influence had effects, among other things, on intellectual property rights. After 1898, the United States promoted a new patent administration in Cuba, Puerto Rico and the Philippines. During the negotiation of the Treaty of Paris (1898) between Spain and the United States, one aspect of the discussion was the regulation of patent and copyright laws. Provisions with regard to patent issues were finally recognised in the Treaty.[76] Cuba became a semi-independent protectorate, neither a fully sovereign republic nor an American colony. This explains why from November 1899 all US patents were recognised in Cuba.[77] For its part, in 1899 Spain gave three months to patentees residing in the colonies to pay their fees if they had not done so during the war.[78] In May 1900, by American military order, it was established that all holders of Spanish colonial patents who wanted protection in the United States should revalidate their patent certificates of registration in front of the new authorities of Cuba.[79] Finally, in 1904, Cuba adhered to the Paris Convention for the Protection of Industrial Property, in addition to having ratified the agreement on patents that was decided at the Second Pan-American Conference held in 1902 in Mexico.

# Notes

1. O. Bracha, *Owning Ideas: The Intellectual Origins of American Intellectual Property* (Cambridge: Cambridge University Press, 2016); L. Bently, 'The "Extraordinary Multiplicity" of Intellectual Property Laws in the British Colonies in the Nineteenth Century', *Theoretical Inquiries in Law* 12 (1), (2011): 161–200.
2. B. Woodcroft, *Alphabetical Index of Patentees of Inventions, 1617–1852* (Reprint: Evelyn, Adams & MacKay, 1969).
3. A. F. Muir, 'Patents and Copyrights in the Republic of Texas', *Journal of Southern History* 12 (1946): 204–222.
4. H. J. Knight, *Confederate Invention: The Story of the Confederate States Patent Office and Its Inventors* (Baton Rouge: Louisiana State University Press, 2011).

5. Bently (2011), Op. cit.; C.A. Nard, 'Legal Forms and the Common Law of Patents', *Boston University Law Review* 90 (51), (2010).
6. R. Sagar, 'Introduction of Exclusive Privileges in Colonial India: Why and for Whose Benefit?', *Intellectual Property Quarterly* 164 (2007).
7. J. Todd, *Colonial Technology: Science and the Transfer of Innovation to Australia* (Cambridge: Cambridge University Press, 1995).
8. S. Ricketson, *The Paris Convention for the Protection of Industrial Property: A Commentary* (Oxford: Oxford University Press, 2015).
9. G. Galvez-Behar, 'Les empires et leurs brevets', in L. Hilaire-Pérez and L. Zakharova (eds.), *Les techniques et la globalisation au XXe siècle* (Rennes: Presses Universitaires de Rennes, 2016).
10. J. Sánchez Gómez, 'Comienzos de la formación de la técnica minometalúrgica colonial', in J. Sánchez Gómez et al. (eds.), *La savia del imperio. Tres estudios de economía colonial* (Salamanca: Universidad de Salamanca, 1997); B. Escobar, 'Doctrines and the Making of an Early Patent System in the Developing World: the Chilean Case. 1840s–1910s', *Documentos de Trabajo de la UDP* 58 (2004); R. Sánchez-Flores, *Historia de la tecnología y la invención en México* (Mexico City: Fomento Cultural Banamex, 1980).
11. J. Coastworth, 'Obstacles to Economic Growth in Nineteenth Century Mexico', *American Historical Review* L. XXXIII (1978): 93.
12. S. J. Patel, 'The Patent System and the Third World', *World Development* 2 (9), (1974): 3–14.
13. E. Beatty, *Institutions and Investment: The Political Basis of Industrialization in Mexico before 1911* (Stanford: Stanford University Press, 2011).
14. Ricketson (2015), Op. cit.
15. Behar (2016), Op. cit.
16. BOPI, year I, no. 7 (01/12/1886).
17. S. Ladas, *Patents, Trademarks, and Related Rights: National and International Protection* (Cambridge, MA: Harvard University Press, 1975); See also *The New York Times* (04/03/1890): 3.
18. 'Tratado sobre patentes de invención, dibujos y modelos industriales, y marcas de comercio y de fábrica. Segunda Conferencia Internacional Americana', México, 22/10/1901–31/01/1902, *Conferencias Internacionales Americanas 1889–1936* (Digital Library Daniel Cosío Villegas, El Colegio de México).
19. C. Schmidt-Nowara, 'La España Ultramarina: Colonialism and Nation-Building in Nineteenth-Century Spain', *European History Quarterly* 34 (2), (2004): 191–214; M. Moreno Fraginals, *Cuba-España, España-Cuba: Historia común* (Barcelona: Crítica, 1995): 168.

20. Royal Cedula 30/07/1833 (BN, Sig. H. A. 17303).
21. N. Fernández de Pinedo, D. Pretel and P. Sáiz, 'Patents, Sugar Technology and Subimperial Institutions in Nineteenth-century Cuba', *History of Technology* 30 (2010): 47–62 and D. Pretel, 'Invenciones institucionales: el sistema de patentes en las colonias españolas durante el siglo XIX', *América Latina en la Historia Económica*, Vol. 26 (2), (2019). For partial studies of patents granted in Cuba see L. Fernández Prieto, *Espacio de poder, ciencia y agricultura en Cuba: el círculo de hacendados, 1878–1917* (Madrid: CSIC, 2008): 113–9; M. A. Marqué Dolz, *Las industrias menores: empresarios y empresas en Cuba, 1880–1920* (Havana: Ciencias Sociales, 2006): 98 and 224.
22. Circular of 31/01/1849 (CLE, t. XLVI).
23. Royal Decree of 30/07/1833, which extended the Royal Decree of 1/03/1826 on privileges of inventions and introductions to overseas domains.
24. On the constitution of the *Junta de Fomento* and its power see L. Marrero, *Cuba: economía y sociedad* (Madrid: Editorial Playor, 1984): 41–2.
25. ANC, Gobierno Superior Civil, Leg. 1478. Exp. 58495.
26. *Memorias de la Sociedad Patriótica* t.IX (Havana: Imprenta del Gobierno y Capitania General, 1839): 167.
27. See, for example, report by Mariano Vieta: ANC, Junta de Fomento, Leg. 95, Exp. 4007.
28. *Memorias de la Sociedad Económica* t.XVII (Havana: Imprenta del Gobierno y Capitania General, 1843): 179–180.
29. *Memorias de la Sociedad Patriótica* t.II (Havana: Imprenta del Gobierno y Capitania General, 1836): 89–99.
30. ANC, records of privileges and patents of invention in the following collections: Gobierno General, Gobierno Superior Civil, Real Consulado, Junta de Fomento e Intendencia de Hacienda.
31. D. Pretel, 'La economía política del sistema español de patentes en perspectiva internacional, 1826–1902', *Investigaciones de Historia Económica* 13 (3), (2017): 190–200; P. Sáiz: 'Transferencia internacional de tecnología hacia España a través del sistema de patentes (1759–1900)', in M. Merger (ed.), *Transferts de technologies en Méditerranée* (Paris: Presses de l'Université Paris-Sorbonne, 2006): 29–52; J. Pella y Forgás, *Las patentes de invención y los derechos del inventor: tratado de utilidad práctica para inventores e industriales* (Barcelona: Administración de Industria e Invenciones, 1892).
32. ANC, Real Consulado y Junta de Fomento, Legajo 95, Exp. 3996 and Exp. 4004.

33. *Memorias de la Real Sociedad Económica* t.ɪ (Havana: Imprenta del Gobierno y Capitania General, 1846): 15–20.
34. F. Knight, 'Origins of Wealth and the Sugar Revolution in Cuba, 1750–1850', *The Hispanic American Historical Review* 57 (2), (1977).
35. *Memorias de la Sociedad Económica* t.xvɪɪ (Havana: Imprenta del Gobierno y Capitania General, 1843): 388–389.
36. ANC, Junta de Fomento, Leg. 95, Exp. 4021 and *Anales de las Reales Junta de Fomento y Sociedad Económica* (Havana, 1849): 293–302.
37. ANC, Real Consulado y Junta de Fomento, Leg. 95, Exp. 3996.
38. D. Rood, *The Reinvention of Atlantic Slavery: Technology, Labor, Race, and Capitalism in the Greater Caribbean* (Oxford: Oxford University Press, 2017); J. Curry-Machado, *Cuban Sugar Industry: Transnational Networks and Engineering Migrants in Mid-Nineteenth Century Cuba* (Basingstoke and New York: Palgrave Macmillan, 2011).
39. D. Pretel and N. Fernández de Pinedo, 'Circuits of Knowledge: Foreign Technology and Transnational Expertise in Nineteenth-century Cuba', in A. Leonard and D. Pretel (eds.), *The Caribbean and the Atlantic World Economy: Circuits of Trade, Money and Knowledge, 1650–1914* (Basingstoke: Palgrave Macmillan, 2015): 263–289; L. Cabrera Salcedo, *De los bueyes al vapor* (San Juan: Editorial UPR, 2010); J. G. Ortega, 'Machines, Modernity, and Sugar: the Greater Caribbean in a Global Context, 1812–50', *Journal of Global History* 9 (1), (2014).
40. J. Alcover, 'Situación agrícola e industrial en la isla de Cuba', *La Gaceta Industrial* 14 (25/07/1884).
41. C. P. Tucker, *Insatiable Appetite: The United States and the Ecological Degradation of the Tropical World* (Berkeley: University of California Press, 2000): 100–109.
42. ANC, Real Consulado, Leg. 204, Exp. 9007 and 9008. See also *Diario de las Actas y Discusiones de las Cortes: legislatura de los años de 1820 y 1821*, Madrid, t. ɪv., pp. 237–246.
43. J. M. Fradera, *Filipinas, la colonia más peculiar* (Madrid: CSIC, 1999): 113, 250 and 277.
44. AHOEPM, Privilegio n° 17.
45. AHOEPM, Privilegio n° 33.
46. J. Macle, 'Los privilegios de invención existentes en el ANC', *unpublished archival presentation* (2013); and P. Sáiz, and N. Fernández de Pinedo, *Base de datos de solicitudes de privilegios de invención. Cuba, 1820–1898* (Madrid: Universidad Autónoma de Madrid, 2014). See also E. Beatty

et al., 'Technology in Latin America's Past and Present: New Evidence from the Patent Records', *Latin American Research Review* 52 (1), (2017). For the case of patents granted in Puerto Rico see L. Cabrera Salcedo, *Inventos para el azúcar. Historia tecnológica puertorriqueña* (Puerto Rico: Siglo XXI-Instituto de Cultura Puertorriqueña, 2007); and Cabrera Salcedo (2010), Op. cit.

47. AHN, Ultramar collection, Ministerio de Ultramar Remite Patentes de Invención, nos. 127, 155, 156, 131 for the years 1888, 1889 and 1891–1893 y AHN, Ultramar, Convalidación de Patentes Dadas en la Península, nos. 178, 180, 182, 184 para los años 1893–1896.

48. *Gaceta de La Habana* (01/01/1851): 1.

49. In the ANC (Correspondencia Fondo Iznaga Valle) there is also documentation of several contracts of foreign machinists employed in the 1840s and 1850s by the Soledad, Niña and Santa Ana sugar mills in the major sugar production area of Trinidad.

50. *Scientific American* 11 (5), (13/10,1855): 38.

51. *Scientific American* 7 (8), (08/11/1851): 59.

52. Curry-Machado (2011), Op. cit.

53. *Scientific American* 1007 (8), (23/08/1862); *Scientific American* 1012 (13), (25/03/1865); and *The London Gazette* (13/01/1863).

54. ANC, Gobierno General, Leg. 441, Exp. 21319.

55. AHOEPM, Privilegio de ultramar No. 85 (11/02/1851); U.S. Patent No. 9087 (29/06/1852).

56. AHOEPM, Privilegio de ultramar No. 50 (22/05/1841).

57. Ortega (2014), Op. cit.

58. ANC, Gobierno Superior Civil, Leg. 1476, Exp. 58365.

59. ANC, Gobierno General, Leg. 333, Exp. 15986.

60. ANC, Gobierno Superior Civil, Leg. 1478, Exp. 58514.

61. Royal Decree of 14/05/1880.

62. F. García Garófalo, *La propiedad intelectual e industrial: Su legislación en la península y provincias ultramarinas* (Havana: La Propaganda Literaria, 1890).

63. G. Vicuña, 'Las patentes y marcas de Ultramar', *La Semana Industrial*, n° 15 (14/04/1882): 144.

64. *Crónica de la Industria*, n° 68, year III (15/10/1877): 293–4.

65. Royal Decree of 21/08/1884, CLE (T. CXXXVI); 'Marcas, modelos y dibujos industriales en Ultramar', *Industria e Invenciones* (06/08/1884): 88–89.

66. Pella (1892), Op. cit.
67. F. Zayas, 'Ingenios Centrales', *Revista de Agricultura*, n° 10 year II (31/10/1880): 310.
68. K. Ferris, 'Technology, Novelty, and Modernity: Spanish Perceptions of the United States in the Late Nineteenth Century', *Hispanic Research Journal* 11 (1), (2010): 37–47.
69. ANC, Gobierno General, Leg. 455, Exp. 22246.
70. Pretel and Fernández (2015), Op. cit.
71. TAEP [XX19KA; TAEM 0:00], 5/09/1881. See also Edison Spanish Colonial Light Co., *La luz Edison / Luz eléctrica incandescente* (New York: Imprenta de N. Ponce de Leon, 1882); W. J. Hausman, et al., *Global Electrification: Multinational Enterprise and International Finance in the History of Light and Power, 1878–2007* (Cambridge: Cambridge University Press, 2008): 77–8.
72. TAEP [D8434ZAI; TAEM 73:826], 08/13/1884.
73. F. L. Dyer and T. C. Martin, *Edison: His Life and Inventions, volume 2* (New York: Harper & Bros., 1910): 971.
74. AHOEPM, Patents Nos. 2984 (13/01/1883); 3112 (17/02/1883); 3122 (17/02/1883).
75. AHOEPM, Patent Nos. 1657 and 2023. See also TAEP [HM820157; TAEM 86:452], 02/09/1882.
76. J. Bellido, et al. 'Commentary on US-Spanish Peace Treaty (1898)', in L. Bently and M. Kretschmer (eds.), *Primary Sources on Copyright, 1450–1900* (www.copyrighthistory.org)
77. Circular of 12/10/1899.
78. Royal Decree of 17/02/1899.
79. Military Order No. 216 (26/05/1900).

# 6

# Inventing Late Industrialisation

**Abstract** This book has addressed many crucial debates on the institutionalisation of patents in nineteenth-century Spain, especially during its
final decades. It has not addressed all such debates or covered all sides of
the debates but has instead concentrated on the aspects of political regulation, bureaucratic organisation, patenting culture and institutional
agency. This epilogue provides an account of the historical relationship
between institutional diversity and political economy. It also discusses the
difference that the patent system made in the technological development
and industrialisation of Spain during the nineteenth century. The final
part of the chapter makes historically grounded claims about the political
economy of patent protection as it pertains to the dynamics of late
industrialisation.

**Keywords** Late industrialisation • Institutional diversity • Political
economy • Technology transfer • Cultural-institutional infrastructure

The history of patents is a history of political economy as much as of
innovation and technological change.[1] Patent rights during the nineteenth

© The Author(s) 2018                                                     **145**
D. Pretel, *Institutionalising Patents in Nineteenth-Century Spain*, Palgrave Studies
in Economic History, https://doi.org/10.1007/978-3-319-96298-6_6

century were an industrial policy that served conflicting interests, ideas and ends. There is little doubt that distinct national cultures and philosophical traditions underlay the varying designs of property laws, economic policies and bureaucratic organisation.[2] However, the dissimilarities among national patent institutions were not only the result of differences in national traditions or legal styles. In the industrialising nineteenth century, patent rights were policy tools used for industrial competition among nations.[3] At the same time, intellectual property rights were pragmatic institutional arrangements that served specific national and international interest groups. The politics of patents reflected the vested interests of the beneficiaries of patent regulation—from inventors to engineers, from companies to intermediaries.

It is clear that, throughout much of the nineteenth century, patent systems took diverse institutional forms in different places. It was not until recently, however, that historians have begun studying this profound diversity in patent cultures.[4] As the historian Ian Inkster notes, whilst the different national patent institutions 'were at least nominally similar and constructed under a set of common understandings, in reality, they were *organised* in such a way as to produce different outcomes'.[5] The differences were not just in legal regulations but also in bureaucratic practices and institutional organisation. The variations in regulation are evident when we look at the *Handbook of Patent Law of All Countries*, a comprehensive study of the day published by the British chartered patent agent William P. Thompson from 1882.[6] The most obvious example of patent diversity is the marked differences that existed between registration systems and systems with novelty examinations. Other disparities can be found in issues such as the composition of concession bodies and the professionalisation of administrative practice; the cost of applications; the status of foreign inventors; the role of intermediaries; the use of working clauses, utility models, patents of importation and compulsory licencing; the duration of monopolies; and patent-eligible subject matters. The central agents working within, through and among these institutions also varied from one country to another. Even when the regulations were almost the same, in some countries national laws were often not enforced and bilateral or international agreements ignored.

The political economy of industrialisation largely explains the variations in patent policies and the adaptative nature of patent institutions during the nineteenth century. The diversity of legal rules, bureaucratic practices and institutional settings was closely related to the degrees of industrialisation and levels of technological dependence across countries. Political conditions likewise affected patent institutionalisation, as colonial cases demonstrate. Much of the diversity among national patent systems can also be explained by various nuances introduced by those who established, reinterpreted and transformed the working rules, which ultimately reflected their own self-interest. Legal variations and the range of institutional approaches make clear that, during the nineteenth century, patent systems were pragmatic and contingent structures that did not observe a pure principle. Such institutional structures were shaped by the historical conjuncture of their invention and development following path-dependence institutional inertia.

In this context, a major challenge is to identify the type of analysis that will yield the best understanding of the institutional complexity of patent systems during the nineteenth century. A quantitative approach to the history of patents—as useful it can be to delineate general trends— obscures the institutional diversity of patent systems. It is not only that patent records are 'archives of failure', to use the expression of historian David Edgerton,[7] but that research based on patent counts tells us little about the why, how, when or who established, shaped and maintained intellectual property rights. These questions can only be answered with a sustained attention to the nuances of the institutional framework and the socio-political circumstances in which patent rights developed.

This book has asserted the need for an institutional study of patent systems. Such research engages in a thick description of patent systems over time, particularly the role of agents and institutional agency. The primary motivation for using this institutionalist perspective as an analytical framework is to avoid a broad economic and cultural determinism wherein institutions and social agency are considered exogenous variables, and instead to bring these more tangible factors back into the core of the analysis of technological systems. The detailed study of institutional nuances and bureaucratic practices can illuminate how economic and socio-political conditions impacted patent systems during the second half of the nineteenth century. For instance, the prospects of late industrialisation during this period mostly explained the institutionalisation of patents in countries at the periphery of industrial development.

The institutional and technological imperatives for late industrialisation are apparent when we consider the making of the Spanish patent system during the nineteenth century. Although there is no clear causality between the introduction of patent protection in Spain and the country's level of industrial innovation, the inverse relationship seems clear. Spain's relative industrial failure during the nineteenth century set the stage for the institutionalisation of its patent system. The Spanish patent system was overtly designed and organised to stimulate inward technology transfer as a response to the perceived backward industrial position of the Spanish economy. The degree of industrial backwardness also explains the patenting culture and broader social infrastructure that supported intellectual property rights in Spain.

Other key features of Spanish patent culture were the country's weak enforcement of the law, its small market for intellectual property rights and its limited role in disseminating the technological information registered at the patent office. In reality, all national systems were, to a greater or lesser extent, laxer in enforcing patent rights than today's systems, often favouring technological emulation of technologies from abroad. However, there is a good deal of evidence to suggest that in the countries of the economic periphery, such as Spain, the inclination to use patent rights to foster technology transfer-in was disproportionately high. For relatively technologically backward countries, patent rights were an instrument by which to acquire foreign technological capacities and potentially to catch up in the process of industrialisation.

At the same time, the study of the Spanish patent system exposes the broader international tendencies of the late nineteenth century. From its conception this institution was rooted in the French model. Spain's institutional dependence on foreign dynamics was also evident in the public political debate and the rhetoric of the specialised press of the day. After a timid controversy during the 1850s when patents came into public scrutiny, Spanish political and economic classes would accept these monopoly rights as non-problematic as long as they served national economic interests. In other words, elites made economic progress a priority and promoted institutional substitution in an effort to catch up to the most advanced industrial countries. Later on, from the 1880s, liberal arguments supporting intellectual property rights—especially from those more affected by their regulation—were reinforced by Spain's relative institutional convergence with other national systems. Even so, intellectual property rights were never at the centre of Spain's public debate throughout the nineteenth century, as they were in other European countries.

It is important to note that the 1880s represented a watershed in the functioning of the Spanish patent system. It was not until that decade that this institution was firmly entrenched. The expansion of the system was associated with the acceleration of a global economy—particularly an increase in the technical flows among countries—and the negotiation and implementation of international agreements regarding intellectual property laws. The weak, peripheral and dependent nature of the system was reinforced through Spain's integration into the Paris Union of 1883. During the 1880s, Spain's patent system became bound up in global patenting networks. Especially relevant was the transnational flow of engineering expertise and the emergence and ubiquity of international intermediaries. Meanwhile, several foreign corporations increasingly patented in Spain during the last two decades of the nineteenth century following a defensive strategy to try to protect their international market power. Under such circumstances, it seems that the lodging of patent claims in Spain did not confer substantial commercial gains on these companies, but rather served to preserve their monopolistic or oligopolistic positions.

In the last two decades of the nineteenth century, a variety of agents placed themselves at the centre of this Spanish institution, becoming influential in its regulation and operation. Engineers, lawyers and other business experts mediated among the realms of patenting, industry and the market, particularly in transnational activities. These experts reduced the risks assumed by foreign patentees while increasing the costs of securing patent rights. They also reduced the time span for lodging rights in this country. In the Spanish case, the generalisation of specialists in patent intermediation was not linked, as it was in other countries, to the expansion of a domestic market for technology, but rather to the explosion of international patenting that occurred during the 1880s and 1890s.

Of particular interest are the colonial patent systems at work in Cuba, Puerto Rico and the Philippines. Despite its imperfection—or precisely because of it—these colonial institutions were thriving as information systems during the mid-nineteenth century. The inconsistency in the granting of patent rights suited the interests of local elites in their collective efforts to promote colonial agrarian economies. In the case of Cuba, landowners in the sugar industry saw the colonial patent system—along with other policies and publications—as a pragmatic collective infrastructure for sharing technological details rather than as an institution to protect inventors. Similarly, additional monopolies and a mix of other rewards were granted to carry out commercial and industrial endeavours. In the case of Spanish colonies, the patent system was not a matter of imperial policy. It was a decentralised institution under the authority of colonial elites with a weak degree of legal coordination with the metropolis. Unlike in the mainland system, in Cuba there was an examination of the novelty and utility of patents. The effects of the Paris Convention of 1883 and of Pan-American Conferences would be felt in the former Spanish colonies after they had already entered the twentieth century. The study of the Cuban patent system makes clear that colonies producing raw materials and basic goods for international markets not only received foreign capital and technologies but also established domestic institutions to organise the circulation of knowledge and resources.

An important question is what difference patent protection made in the industrialisation and economic development of modern Spain. There is no empirical evidence that Spanish patent regulations increased patterns of innovation between 1826 and 1902. The outcomes of Spain's patent institution were quite modest, but also modest was the level of Spanish industrialisation and technological capacities relative to industrial leaders and successful latecomers. The role of patents as an incentive for invention was minimal due to the timid nature of Spanish industrialisation but due also to how the system operated. Patents, however, may have favoured technology transfer and technological imitation. The very fact that many local industrialists obtained patents of introduction indicates that they mattered. The patent system may have been a mechanism of technological knowledge diffusion, but it was clearly not the only one nor was it strictly necessary.

The Spanish case illustrates that industrialisation was not easily planned. If anything, the low levels of patenting in Spain and the precise characteristics of the Spanish system (e.g. its bureaucratic framework, market for patents and legal enforcement) speak to the failure of patents as an effective industrial policy for late industrialisation. However, the quantitative data on patenting activity and the study of the changing nature of patent culture alone cannot answer the question of whether the Spanish patent system was a successful policy or institutional arrangement. Such a question entails looking at alternative technological policies in a difficult exercise of counterfactual history that looks at what would have happened if patent rights had never been established in Spain in the first place. Moreover, a definitive balance sheet of the overall impact of the patent system in modern Spain would depend on the criteria adopted to assess this institution, whether as a policy for economic growth or as an arrangement for the wider dissemination of useful technological knowledge. The patent system understood beyond its economic role still needs to be considered more fully.

A different question has to do with the impact of new technologies in the Spanish society and economy. Such a study would require a focus on the actual technologies-in-use in specific production sites and not only an analysis of patenting data and culture. The information contained in patents was neither the only nor the most important source of knowledge for the development of specific technologies-in-use in Spanish industry and agriculture. The granting of a patent in Spain was only evidence that someone had met the legal and administrative requirements of patentability, not necessarily of successful industrial innovation or technology transfer. Foreign technologies were central not just in the take-off of Spain's textile and metallurgical industries but also in the expansion of its transport network and infrastructures for the distribution of gas and electricity. However, successful adoption of foreign technologies through imports did not mean that economic convergence with other European countries had been achieved, nor that these sectors were competitive in international markets. More important for the content of this book, it seems clear that most of the cases of successful transfer of foreign technologies to Spain were not channelled through the patent system but indirectly through trade. The levels of technology imports remained high and most of the industrial technologies patented in Spain were not manufactured in that country—an example being railway technologies. As a result, higher levels of patenting between 1878 and 1902 did not reduce the technological gap with industrial leaders and did not necessarily translate into industrialisation. Spain's technological backwardness would continue to be evident at international and national industrial exhibitions at the turn of the twentieth century.

If anything, the Spanish patent system, as an information system, created institutional facilities for the development of dense networks for the exchange and advertisement of industrial technologies. Its relative success as an information system was connected with its openness to technological, institutional and ideological influences from outside. The Spanish patent office (with its changing names and various administrative structures) engendered the rise of a community of skilled professionals with expertise in technological and industrial questions. Another benefit was the rise of a wider cultural-institutional infrastructure around this system—an infrastructure composed of specialised presses, museums, technological exhibitions and engineering schools. The formal disclosure of information to Spanish industrial classes could promote innovation (often through imitation) even when the law was weak in enforcing intellectual property rights. This public dimension of science and technology, that consolidated in Spain from the 1870s, had an importance that transcended the history of patents.

The lessons stemming from the institutionalisation of the Spanish patent system in the nineteenth century may well be of broader significance. During the second half of the nineteenth century, institutional reform could promote technological progress and the overcoming of technological dependency. The construction of the Spanish patent system was marked by the determinants of the international economy, following the institutional adaptation scheme proposed by Alexander Gerschenkron for countries with relative industrial backwardness.[8] Technological backwardness underscores the imperative of institutional innovation—specifically among institutions governing property rights—to incentivise catching up. The new industrial paradigm of the age of 'machinofacture' transformed the barriers and imperatives for late industrialisation.[9] For instance, late industrialisation required increasing participation by the state in economic activity and higher investments in formal technical education, public infrastructure and technological systems.

Over the course of the nineteenth century, technology transfer from early adopters presented an opportunity for latecomers to industrialise, particularly in the years 1870–1900. However, imitation and adoption were not always easy. Technologies and organisational processes were increasingly complex and expensive and thus difficult to transfer. Technological innovation could be inhibited by a variety of factors, from economic (demand, the cost of new technology) to social (attitudes, social relations) and from political (institutional responses, instability) to spatial (social distances, population density).[10] Technology transfer also depended on the cost of the transfer of reliable technical information and the domestic capacity to adapt and use the new technologies[11] Technology transfer could generate a learning process, but at the same time the dependence that often ensued from technology transfer to latecomers could inhibit the development of local expertise.

Late industrialisation must be understood within the international context of both the great industrial divide and the race for technologies among nations that was occurring at the time. The late nineteenth-century Spanish patent system was the consequence of the historical making of an international patent system in which different nation-states had distinctive roles. It was an international patent system built on the growing interdependencies among national economies. The level of patenting for each national system correlated with its national patterns of production and trade. The unbalanced industrial structure of the world economy explains the asymmetric level of patents issued around the world. As most scholars have pointed out, the overwhelming majority of patents issued worldwide until 1914 were concentrated in just a few countries (France, Britain, Germany and the United States). Other countries such as Spain had weak, unstable, pragmatic and small patent systems.

From the mid-nineteenth century patent systems became an increasingly hegemonic institution at the international level—a development with implications for the political economy of industrialisation of various countries, particularly in Europe and the Americas. In the period of industrial catch-up that characterised the late nineteenth century, institutions adapted to the new economic environment and the process of capitalist globalisation. Modernising elites in peripheral economies emulated the institutional structure of advanced industrial nations. But the transplanted institutions were adapted to local conditions and organised differently. Peripheral-dependent nations created patent systems that focused less on providing incentives for indigenous inventors than on stimulating the diffusion of foreign technologies.

The prevailing level of industrial development and the late nineteenth century's uneven technological progress across nations explain the organisation and functioning of peripheral patent systems. Even so, in the case of Spain, state-induced monopoly rights had enjoyed little success as an instrument of technological innovation, showing that nineteenth-century industrialisation was not only a matter of proper regulation by the state, as industrialisation could be inhibited by a variety of broader underlying socio-economic processes and technological imperatives. From a political economy perspective, it seems that nineteenth-century Spain was industrially backward not so much because of the anomalousness of its culture or institutions, but rather because of global barriers and imperatives inherent to the paradigm of late technological development.

# Notes

1. C. May, *A Global Political Economy of Intellectual Property Rights. The New Enclosures?* (London: Routledge, 2000); J. Mokyr, 'The Political Economy of Technological Change: Resistance and Innovation in Economic History', in M. Berg and K. Bruland (eds.), *Technological Revolutions in Europe* (Cheltenham: Edward Elgar Publishers, 1998): 39–64; P. A. David, 'Intellectual Property Institutions and the Panda's Thumb: Patents, Copyrights, and Trade Secrets in Economic Theory and History', in M. Wallerstein et al. (eds.), *Global Dimensions of Intellectual Property rights in Science and Technology* (Washington, D.C.: Nacional Academy Press, 1993): 19–61.
2. For a discussion on the relevance of national cultures in industrial policy and economic regulation see F. Dobbin, *Forging Industrial Policy: The United States, Britain, and France in the Railway Age* (Cambridge: Cambridge University Press, 1994).

3. H-J. Chang, 'Intellectual Property Rights and Economic Development', *Journal of Human Development*, 2(2), (2001): 287–309.
4. S. Arapostathis and G. Gooday, *Patently Contestable: Electrical Technologies and Inventor Identities on Trial in Britain* (Cambridge, MA: MIT University Press, 2013); James F. Stark, 'Introduction: Plurality in Patenting: Medical Technology and Cultures of Protection', *The British Journal for the History of Science* 49 (4), (2016): 533–540; M. Biagioli et al. (eds.), *Making and Unmaking Intellectual Property* (Chicago: University of Chicago Press, 2011); S. Wilf and G. Gooday (eds.), *International Diversity in Patent Cultures* (Cambridge: Cambridge University Press, forthcoming).
5. I. Inkster, 'Patent Agency: Problems and Perspectives', *History of Technology* 31 (2012): 89–97.
6. W. P. Thompson, *Handbook of Patent Law of All Countries* (London: Stevens & Sons, 1882).
7. D. Edgerton, 'The Political Economy of Science – prospects and retrospects' in D. Tyfield et al. (eds.), *The Routledge Handbook of the Political Economy of Science* (London: Routledge, 2017): 21–31.
8. A. Gerschenkron, *Economic Backwardness in Historical Perspective* (Cambridge, MA: Belknap Press of Harvard University Press, 1962). See also I. Inkster, 'Politicising the Gerschenkron Schema: Technology Transfer, Late Development and the State in Historical Perspective', *The Journal of European Economic History* 31 (1), (2002): 45–87.
9. For the historical transitions between different socio-technical systems and the linkages between technology and economy see C. Freeman and F. Louçã, *As Time Goes By: From the Industrial Revolutions to the Information Revolution* (Oxford: Oxford University Press, 2001): 140–3; C. Pérez, 'Technological Revolutions, Paradigm Shifts and Socio-Institutional Change', in E. S. Reiner (ed.), *Globalization, Economic Development and Inequality: An Alternative Perspective* (Cheltenham: Edward Elgar, 2004): 217–42.
10. N. Von Tunzelmann, *Technology and Industrial Progress: The Foundations of Economic Growth* (Cheltenham: Edward Elgar, 1995): especially pp. 7–10 and chapter 5.
11. H. P. Binswanger and V. W. Ruttan, *Induced Innovation: Technology, Institutions, and Development* (Baltimore: Johns Hopkins University Press, 1978); I. Inkster, 'Motivation and Achievement: Technological Change and Creative Response in Comparative Industrial History', *Journal of European Economic History* 27 (1998): 29–66; N. Rosenberg, 'Factors Affecting the Diffusion of Technology', *Explorations in Economic History* 10 (1972): 3–33.

# Appendix: Notes on Sources

This study of the institutionalisation of patents in nineteenth-century Spain derives from various primary and secondary sources. Many of the arguments presented in this book arose from a detailed study of original patent documentation in Spanish archives. The contentions of this book rest on evidence gleaned from thousands of original patent files for the period 1826–1902, all located in the Spanish Patent and Trademark Office Historical Archive (AHOEPM). Additionally, I have made extensive use of the historical database on patents of invention established by the Autonomous University of Madrid and the Spanish Patent and Trademark Office (OEPM). A research team under the direction of Professor Patricio Sáiz created this database, in whose elaboration and study I participated as a researcher.[1]

For the purpose of this book, I built two databases of patent agents in the Spanish system through the study of the original powers of attorney kept in the patent documentation of the AHOEPM. The first database consists of all patent applications and assignments (transfers) filed between 1826 and 1878; the second consists of patent applications and assignments for the years 1880, 1890 and 1900.

© The Author(s) 2018
D. Pretel, *Institutionalising Patents in Nineteenth-Century Spain*, Palgrave Studies in Economic History, https://doi.org/10.1007/978-3-319-96298-6

In order to understand the functioning of the colonial patent system and patenting activities in the Spanish colonies during the nineteenth century, I also consulted the original files on patents and related documentation at the Cuban National Archive (ANC) in Havana.

Agents left few explicit historical private records of their activities, but I was fortunate enough to obtain permission to consult the original business diaries and registration books of the lawyer Julio Vizcarrondo for the period 1875–1888—records conserved at the private archive of the industrial property firm Elzaburu in Madrid (ELZ).

In an effort to provide broader insights into the functioning of the Spanish system, I have consulted numerous supplementary sources such as technical presses, official publications, transactions of professional associations, commercial directories, business material, parliamentary debates and contemporary legislation. Particularly worthy of mention are the publications *Industria e Invenciones* (Barcelona), *La Gaceta Industrial* (Madrid), *Transactions of the Chartered Institute of Patent Agents* (London) and *Scientific American* (New York). Also relevant for the content of this book are the official publications *Boletín Oficial de la Propiedad Industrial* (BOPI, Madrid) and *Memorias de la Sociedad Económica de Amigos del País* (Havana).

# Recommended Further Reading

Abbott, A., *The System of Professions: An Essay on the Division of Expert Labor* (Chicago: University of Chicago Press, 1988).

Amengual, R., *Bielas y Alabes, evolución histórica de las primeras máquinas térmicas a través de las patentes españolas, 1826–1914* (Madrid: OEPM, 2008).

Beatty, E. et al., 'Technology in Latin America's Past and Present: New Evidence from the Patent Records', *Latin American Research Review* 52 (1), (2017).

Bently, L., 'The "Extraordinary Multiplicity" of Intellectual Property Laws in the British Colonies in the Nineteenth Century', *Theoretical Inquiries in Law* 12 (1), (2011): 161–200.

Biagioli, M. et al. (eds.), *Making and Unmaking Intellectual Property* (Chicago: University of Chicago Press, 2011).

Boldrin, M., *Against Intellectual Monopoly* (Cambridge: Cambridge University Press, 2008).

Bowker, G., 'What's in a Patent?', in W. E. Bijker and J. Law (eds.), *Shaping Technology Building Society. Studies in Sociotechnical Change* (Cambridge, MA: MIT Press, 1992): 53–74.

Bracha, O., *Owning Ideas: The Intellectual Origins of American Intellectual Property* (Cambridge: Cambridge University Press, 2016).

Chang, J-H., 'Institutions and Economic Development: Theory, Policy and History', *Journal of Institutional Economics* 7 (4), (2011): 473–498.

© The Author(s) 2018

D. Pretel, *Institutionalising Patents in Nineteenth-Century Spain*, Palgrave Studies in Economic History, https://doi.org/10.1007/978-3-319-96298-6

Daunton, M. J. and F. Trentmann (eds.), *Worlds of Political Economy: Knowledge and Power in the Nineteenth and Twentieth Centuries* (Basingstoke: Palgrave Macmillan, 2004).

David, P. A., 'Intellectual Property Institutions and the Panda's Thumb: Patents, Copyrights, and Trade Secrets in Economic Theory and History', in M. B. Wallerstein et al. (eds.), *Global Dimensions of Intellectual Property rights in Science and Technology* (Washington, D.C.: Nacional Academy Press): 19–61.

Edgerton, D., 'The Political Economy of Science – Prospects and Retrospects', in D. Tyfield et al. (eds.), *The Routledge Handbook of the Political Economy of Science* (London: Routledge, 2017): 21–31.

Fox, R. and A. Guagnini, *Laboratories, Workshops, and Sites: Concepts and Practices of Research in Industrial Europe, 1800–1914* (Berkeley: University of California Press, 1999).

Freeman, C. and F. Louçã, *As Time Goes By: From the Industrial Revolutions to the Information Revolution* (Oxford: Oxford University Press, 2001).

Gooday, G. and S. Arapostathis, *Patently Contestable: Electrical Technologies and Inventor Identities on Trial in Britain* (Cambridge, MA: MIT University Press, 2013).

Galvez-Behar, G., 'Les Empires et leurs brevets', in L. Hilaire-Pérez and L. Zakharova (eds.), *Les techniques et la globalisation au XXe siècle* (Rennes: Presses Universitaires de Rennes, 2016): 281–296.

Galvez-Behar, G., *La République des Inventeurs: Propriété et Organisation de l'Innovation en France, 1791–1922* (Rennes: Presses Universitaires de Rennes, 2008).

García Tapia, N., *Patentes de invención españolas en el siglo de oro* (Madrid: OEPM, 1994).

Gavroglu, K. et al., 'Science and Technology in the European Periphery: Some Historiographical Reflections', *History of Science* 46 (2), (2008): 153–175.

Gerschenkron A., *Economic Backwardness in Historical Perspective* (Cambridge, MA: Belknap Press of Harvard University Press, 1962).

Inkster, I., 'Patents as Indicators of Technological Change and Innovation: An historical Analysis of the Patent Data 1830–1914', *Transactions of the Newcomen Society*, 73(2), (2003): 179–208.

Inkster, I., 'Technology in World History: Cultures of Constraint and Innovation, Emulation, and Technology Transfers', *Comparative Technology Transfer and Society* 5 (2), (2007): 108–127.

Inkster, I., 'Patent Agency: Problems and Perspectives', *History of Technology* 31 (2012): 89–97.

Jaffe, A. B. and J. Lerner, *Innovation and its Discontents* (Princeton: Princeton University Press, 2004).

Jeremy D. J. (ed.), *International Technology Transfer. Europe, Japan and the USA, 1700–1914* (Aldershot: Edward Elgar, 1991).

Khan, Z. B., 'Selling Ideas: an International Perspective on Patenting and Markets for Technological Innovations, 1790–1930', *Business History Review* 87 (2013): 39–68.

Kranakis, E., 'Patents and Power: European Patent-System Integration in the Context of Globalization', *Technology and Culture* 48 (4), (2007): 689–728.

Ladas, S., *Patents, Trademarks, and Related Rights: National and International Protection* (Cambridge, MA: Harvard University Press, 1975).

Lafuente, A. et al. (eds.), *Maquinismo ibérico* (Madrid: Doce Calles, 2007).

López, S. and J. M. Valdaliso (eds.), *¿Qué inventen ellos?: Tecnología, empresa y cambio económico en la España contemporánea* (Madrid: Alianza Editorial, 1997).

Machlup, F. and E. Penrose, 'The Patent Controversy in the Nineteenth Century', *The Journal of Economic History* 10 (1), (1950): 1–29.

May, C. and S. K. Shell, *Intellectual Property Rights: A Critical History* (London: Lynee Rienner Publishers, 2006).

Mokyr, J., 'The Political Economy of Technological Change: Resistance and Innovation in Economic History', in M. Berg and K. Bruland (eds.), *Technological Revolutions in Europe* (Cheltenham: Edward Elgar Publishers, 1998): 39–64.

North, D. C., 'A Recommendation on How to Intelligently Approach Emerging Problems in Intellectual Property Systems', *Review of Law and Economics* 5 (3), (2009): 1131–1133.

North, D. C., *Institutions, Institutional Change and Economic Performance* (Cambridge: Cambridge University Press, 1990).

Ortiz-Villajos, J. M., 'Spanish Patenting and Technological Dependency, pre-1936', *History of Technology* 24 (2002): 203–32.

Ortiz-Villajos, J. M., *Tecnología y desarrollo económico en la historia contemporánea: Estudio de las patentes registradas en España entre 1882 y 1935* (Madrid: OEPM, 1999).

Ostrom, E., *Understanding Institutional Diversity* (Princeton: Princeton University Press, 2005).

Papanelopoulou, F. et al. (eds.), *Popularizing Science and Technology in the European Periphery, 1800–2000* (Aldershot: Ashgate, 2009).

Patel, S. J., 'The Patent System and the Third World', *World Development* 2 (9), (1974): 3–14.

Pérez, C., 'Technological Revolutions, Paradigm Shifts and Socio-Institutional Change', in E. S. Reiner (ed.), *Globalization, Economic Development and Inequality: An Alternative Perspective* (Cheltenham: Edward Elgar, 2004): 217–42.

Pinch, T., 'Technology and Institutions: Living in a Material World', *Theory and Society* 37 (5), (2008): 461–483.

Plasseraud, Y. and F. Savignon, *Paris 1883: Genèse du Droit Unioniste des Brevets* (Paris: Litec, 1983).

Pollard, S., *Peaceful Conquest: The Industrialization of Europe 1760–1970* (Oxford: Oxford University Press, 1981).

Prados de la Escosura, L., *Spanish Economic Growth, 1850–2015* (London: Palgrave Macmillan, 2017).

Pretel, D., 'El sistema de patentes en las colonias españolas durante el siglo XIX', *América Latina en la Historia Económica*, Vol. 26 (2), (2019).

Pretel, D. and L. Camprubí, *Technology and Globalisation: Networks of Experts in World History* (London: Palgrave Macmillan, 2018).

Pretel, D., 'Invención, nacionalismo tecnológico y progreso: el discurso de la propiedad industrial en la España del siglo XIX', *Empiria*, 18 (2009): 59–83.

Pretel, D., 'La economía política del sistema español de patentes en perspectiva internacional, 1826–1902', *Investigaciones de Historia Económica* 13 (3), (2017): 190–200.

Pretel, D. and N. Fernández de Pinedo, 'Circuits of Knowledge: Foreign Technology and Transnational Expertise in Nineteenth-century Cuba', in A. Leonard and D. Pretel (eds.), *The Caribbean and the Atlantic World Economy: Circuits of trade, money and knowledge, 1650–1914* (Basingstoke: Palgrave Macmillan, 2015): 263–289.

Pretel, D. and P. Sáiz, 'Patent Agents in the European Periphery: Spain (1826–1902)', *History of Technology* 31 (2012): 97–114.

Ricketson, S., *The Paris Convention for the Protection of Industrial Property: A Commentary* (Oxford: Oxford University Press, 2015).

Ringrose, D. R., *Spain, Europe, and the 'Spanish Miracle', 1700–1900* (Cambridge: Cambridge University Press, 1998).

Rosenberg, N., *Perspectives in Technology* (Cambridge: Cambridge University Press, 1976).

Shiva, V., *Patents: Myths and Reality* (New Delhi: Penguin Books, 2001).

Sáiz, P. and D. Pretel, 'Why Did Multinationals Patent in Spain? Several Historical Inquiries', in P-Y. Donzé and S. Nishimura (eds.), *Organizing Global Technology Flows: Institutions, Actors, and Processes* (New York: Routledge, 2013): 39–59.

Sáiz, P., 'Did Patents of Introduction Encourage Technology Transfer? Long-term Evidence from the Spanish Innovation System', *Cliometrica* 8 (1), (2014): 49–78.

Sáiz, P., 'The Spanish Patent System (1770–1907)', *History of Technology* 24 (2002): 45–79.

Sáiz, P., *Invención, patentes e innovación en la España contemporánea* (Madrid: OEPM, 1999).

Streb, J., 'The Cliometric Study of Innovations', in C. Diebolt and M. Haupert (eds.), *Handbook of Cliometrics* (Berlin: Springer Reference, 2016): 447–468.

Swanson, K., 'The Emergence of the Professional Patent Practitioner', *Technology and Culture* 50 (3), (2009): 519–548.

Todd, J., *Colonial Technology: Science and the Transfer of Innovation to Australia* (Cambridge: Cambridge University Press, 1995).

Tortella, G., *El desarrollo de la España contemporánea: Historia económica de los siglos XIX y XX* (Madrid: Alianza Editorial, 2004).

Von Tunzelmann, N., *Technology and Industrial Progress: The Foundations of Economic Growth* (Cheltenham: Edward Elgar, 1995).

# Notes

1. P. Sáiz and F. Cayón (dirs.), *Base de datos de solicitudes de patentes de invención (1878–1939)* and P. Sáiz, *Base de datos de solicitudes de privilegios de invención (1826–1878)*. http://historico.oepm.es

# Index[1]

[1] Note: Page numbers followed by 'n' refer to notes.

© The Author(s) 2018
D. Pretel, *Institutionalising Patents in Nineteenth-Century Spain*, Palgrave Studies in Economic History, https://doi.org/10.1007/978-3-319-96298-6